Sustainability in Coffee Production

T0300299

Coffee, as a commodity and through its global value chains, is the focus of much interest to achieve fair trade and equitable outcomes for producers, processors and consumers. It has iconic cultural and economic significance for Colombia, which is one of the world's major coffee producers for the global market. This book examines sustainable coffee production in Colombia, specifically the initiatives of Nestlé to create shared value.

It describes the transformation of the coffee landscape by the development of economically, socially and environmentally viable and dedicated supply chains. Suppliers have been encouraged to shift production and quality paradigms, in order to develop long-term and sustainable strategies for higher value and premium quality products. This has been partially achieved by establishing a robust partnership with the Coffee Growers Federation and other public, private and social actors, thereby taking control of the institutional architecture and knowledge base that exists in the country. The book provides an important lesson of corporate social responsibility and the creation of shared value for the benefit of farmers, corporations and consumers.

Andrea Biswas-Tortajada is a Research Fellow at the Third World Centre for Water Management in Mexico. She has also worked for the Water Supply Department in Gujarat, India and for the United Nations Development Programme in Geneva.

Asit K. Biswas is Distinguished Visiting Professor at the Lee Kuan Yew School of Public Policy, National University of Singapore. A recipient of the Stockholm Water Prize and named by Reuters as one of the top 10 water trailblazers of the world, his work has been translated into 37 languages.

Sustainability in Coffee Production

Creating Shared Value Chains in Colombia

Andrea Biswas-Tortajada and Asit K. Biswas

Routledge
Taylor & Francis Group

LONDON AND NEW YORK

First published 2015
by Routledge

2 Park Square, Milton Park, Abingdon, Oxfordshire OX14 4RN
711 Third Avenue, New York, NY 10017

Routledge is an imprint of the Taylor & Francis Group, an informa business

First issued in paperback 2017

British Library Cataloguing in Publication Data
A catalogue record for this book is available from the British Library

Library of Congress Cataloging in Publication Data
Biswas-Tortajada, Andrea.
Sustainability in coffee production : creating shared value chains in
Colombia / Andrea Biswas-Tortajada and Asit K. Biswas.
pages cm
Includes bibliographical references and index.
1. Coffee industry--Social aspects--Colombia. 2. Coffee industry--
Environmental aspects--Colombia. 3. Nestl?--Management. 4. Social
responsibility of business--Colombia. 5. Sustainable agriculture--Colombia.
I. Biswas, Asit K. II. Title.
HD9199.C62B57 2015
338.7'6337309861--dc23
2014046579

ISBN: 978-1-138-90207-7 (hbk)
ISBN: 978-0-8153-8163-1 (pbk)

Typeset in Bembo by
Saxon Graphics Ltd, Derby

Contents

Acronyms, abbreviations and definitions

ABACO	Asociación de Banco de Alimentos de Colombia
ACDI	Agricultural Cooperative Development International
ADAM	Desarrollo Alternativo Municipal
ADB	Asian Development Bank
AECID	Agencia Española de Cooperación Internacional para el Desarrollo
AES	Agricultural Extension Services
AIS	Agro Ingreso Seguro
Anacafé	Asociación Nacional de Café (Guatemala)
ANDACOL	Asociación Nacional de Anunciantes de Colombia
ANDI	Asociación Nacional de Empresarios de Colombia
ASI	Aluminium Stewardship Initiative
B2B	Business-to-business
BAP	Best Agricultural Practices
BEPs	Beneficios Económicos Periódicos
C.A.F.E.	Coffee and Farmer Equity
CATIE	Centro Agronómico Tropical de Investigación y Enseñanza
CCL	Colombian Coffee Landscape
CCLC	Coffee Cultural Landscape of Colombia
CCT	Cámara de Comercio de Tuluá
CDP	Carbon Disclosure Project
CDPW	Carbon Disclosure Project Water
CECODES	Consejo Empresarial Colombiano para el Desarrollo Sostenible
CEDE	Center of Economic Development Studies
CEF	Corporate Economic Forum
CELAA	Club du Recyclage des Emballages Légers en Aluminium et Acier
Cenicafé	Centro Nacional de Investigaciones del Café (Colombia)
CEO	Chief Executive Officer
CID	Centro de Investigación y Desarrollo, Universidad Nacional de Colombia
CIMS	Centro de Inteligencia sobre Mercados Sostenibles

CINARA	Instituto de Investigación y Desarrollo en Abastecimiento de Agua, Saneamiento Ambiental y Conservación del Recurso Hídrico
CODHES	Consultoría para los Derechos Humanos y el Desplazamiento
COP	Colombian peso
COSA	Committee on Sustainability Assessment
CPMC	Consejos Participativos de Mujeres Cafeteras
CRECE	Centro de Estudios Regionales Cafeteros y Empresariales
CSR	Corporate Social Responsibility
CSV	Creating Shared Value
CUT	Central Unitaria de Trabajadores de Colombia
DEA	Data Envelopment Analysis
DJSI	Dow Jones Sustainability Indices
DPA	Dairy Partners of Americas
DPN	Departamento de Planeación Nacional
DRE	Programa Desarrollo Rural con Equidad
DSE	Driving Sustainable Economies
ECLAC	United Nations Economic Commission for Latin America/ Comisión Económica para América Latina (CEPAL)
ESG	Environmental, social and corporate governance
Expocafé®	Sociedad Exportadora de Café de las Cooperativas de Caficultores SA
FARC	Fuerzas Armadas Revolucionarias de Colombia
FC	Fondo Nacional del Café
FDI	Foreign Direct Investment
Fedesarrollo	Fundación para la Educación Superior y el Desarrollo
Finagro	Fondo para el Financiamiento del Sector Agropecuario
FLO	Fairtrade Labelling Organisation
FNC	Federación Nacional de Cafeteros de Colombia
FOSYGA	Fondo de Solidaridad y Garantía
FSG	Foundation Strategy Group
GIZ	German Society for International Cooperation
GTZ	German Technical Cooperation
IBLF	International Business Leaders Forum
ICA	Instituto Colombiano Agropecuario
ICA	International Coffee Agreement
ICO	International Coffee Organisation
ICOC	Índice de Conflicto en Colombia
ICR	Incentivo a la Capitalización Rural
IDB	Inter-American Development Bank/Banco Inter-Americano de Desarrollo (BID)
IDPs	Internally Displaced Persons
IDS	Institute for Development Studies
IFC	International Finance Corporation
IGAC	Instituto Geográfico Agustín Codazzi

IISD	International Institute for Sustainable Development
ILO	International Labour Organisation
INCAE	Instituto Centroamericano de Administración de Empresas
INVIAS	Instituto Nacional de Vías
ITC	International Trade Centre
IUCN	International Union for Conservation of Nature
MANA	Mejoramiento Alimenticio y Nutricional de Antioquia
MNC	Multinational Corporation
NSAB	Nespresso Sustainability Advisory Board
OCHA	United Nations Office for the Coordination of Humanitarian Affairs
OECD	Organisation for Economic Co-operation and Development
PROMECAFE	Programa Cooperativo Regional para el Desarrollo Tecnológico y Modernización de la Caficultura
RA	Rainforest Alliance
REPIC	Renewable Energy & Energy Efficiency Promotion in International Cooperation
SAN	Sustainable Agriculture Network
SDC	Swiss Agency for Development and Cooperation
SENA	Servicio Nacional de Aprendizaje
Sinaltrainal	Sindicato Nacional de Trabajadores de la Industria de Alimentos
Sintraimagra	Sindicato Nacional de Trabajadores de la Industria de Productos Grasos y Alimenticios
SMEs	Small and Medium Enterprises
TASQ™	Tool for the Assessment of Sustainable Quality
UN	United Nations
UNDP	United Nations Development Programme
UNEP	United Nations Environment Programme
UNESCO	United Nations Educational, Scientific and Cultural Organization
UNIDO	United Nations Industrial Development Organization
USAID	United States Agency for International Development
VOCA	Volunteers in Overseas Cooperative Assistance
WBCSD	World Business Council for Sustainable Development
WEF	World Economic Forum
WTO	World Trade Organization

Weights and measures

1 bag of coffee = 60 kilograms = 132.3 pounds
1 hectare = 10,000 m^2
1 ton = 16.67 bags
1 ton = 1,000 kilograms

Figures and tables

Figures

Tables

Foreword

Shared value from coffee

Juan Manuel Santos, President of Colombia

As President of the Republic and long-time friend of the coffee guild, it is very satisfying to know the work Nestlé – and its brands Nespresso and Nescafé – has done in recent years to create shared value in the communities it sources from and in its Bugalagrande factory in Valle del Cauca.

Shared value is a business model that creates value for both the great entrepreneur and the producer. We have promoted this throughout this government.

Prosperity for everyone, for which we work tirelessly, has to be prosperity with everyone and for that, the private sector is an invaluable ally who accompanies us in order to build a Colombia with opportunities.

From different fronts, strategic partnerships with foundations, banks, construction companies, NGOs, utilities and the private sector, among others, have enabled us to improve the quality of life of the needy and most vulnerable Colombians.

So, in the past four years we fulfilled the goal of creating 2.5 million jobs, we holistically treated more than one million children with the 'Zero to Always' (*Cero a Siempre*)[1] strategy; we structured and began to implement the most ambitious road infrastructure plan in our history; and we built or started to build more than 900,000 homes, of which 100,000 were given out for free.

This is just to mention some of the projects in which employers are contributing significantly to the realisation of a Colombia that grows fairly by activating a virtuous circle of opportunity.

Our government is determined to promote rural development in our country as never before. We are building, along with our farmers – including coffee growers – a comprehensive agricultural policy with which we hope our fields begin to recover from decades of neglect.

The coffee sector plays a vital role in the national economy and that is why we support it through the Rural Training Incentive (ICR) and the Coffee Farmer Income Protection (CIP) scheme, thanks to which we managed to increase annual production to 11.3 million bags of coffee.

Of course, we still have a long way to go and what better way than to move forward alongside companies conscious of their transformative role in society.

Nestlé is the largest buyer of coffee – and fifth largest buyer of milk – in Colombia. Its policy of shared value creates a great impact on the improvement of production processes, the welfare of its suppliers as well as on the creation of environmentally sustainable production systems.

This publication, which reflects the vision we share with Nestlé, will allow the reader to know the many challenges faced by this leading company that not only took a gamble to build an industry in our country but that is also determined to progress hand in hand the communities it works with.

I appreciate the support the National Coffee Growers' Federation has given to this publication that serves as a stimulus and guide for more companies to make the decision to promote their own development and that of the society by creating shared value.

'No one can be rich if his neighbours are poor,' said John F. Kennedy. This is understood by companies like Nestlé, who know that the only worthy progress is the one that is shared with everyone.

Note

1   *Cero a Siempre* channels the efforts of public, private, social and international actors into a National Strategy for Integral Attention to Early Childhood. This comprehensive framework compiles previously isolated public policies, programmes, services, initiatives and specific actions targeted at improving health, nutrition and education of infants and children up to five years of age (Presidential Council for Early Childhood, 2014).

Preface

Coffee, the focus of this study, is seen as one of the world's most powerful tools for social change (Rice and McLean, 1999 in Linton, 2005). This is perhaps due to its extensive and global value chain, or because it represents a textbook case of a commodity exclusively produced in developing countries and predominantly consumed in affluent societies. In fact, coffee is a highly concentrated sector. Demand comes almost entirely from only 26 countries and almost half of the sector's business activity is controlled by five roasters: Nestlé, Kraft, Procter & Gamble, Sara Lee and Tchibo. In contrast, coffee is produced in over 80 countries and is the livelihood of more than 25 million farmers and over 100 million workers spread all along the supply and value chain.

Coffee has also become one of the most popular items through which consumers and companies alike are expecting and looking to achieve, at times, seemingly exclusive and hard to verify goals: social justice, economic prosperity and environmental sustainability. It has, in fact, become a totemic issue for trade enthusiasts eager to show how a more sustainable 'coffee trade helps spread prosperity worldwide' (Global Envision, 2005). It is also a commodity where the needs of citizens, producing communities, shareholders and consumers are carrying increasingly equal weights even when the vast majority of the market is controlled by a handful of demanding countries and only five major transnational corporations.

It is logical, and yet not always totally instinctive, that the vast amount of raw materials the company requires to feed its global production processes come from an extensive network of thousands of small farmers in procurement markets. Invariably, factors affecting these seemingly distant and scattered supplying communities have an impact on a company's supply and value chain. Take for instance Nestlé, the inventor of instant coffee and of the portioned coffee capsule system. This company is one of the big corporations controlling the international coffee value chain. It is also the world's largest coffee purveyor.

In 2013, it purchased 148,198 tonnes of green coffee directly from farmers, co-operatives or selected traders. That figure includes 15 per cent of Colombia's coffee production, which makes the firm the largest coffee buyer in that country. Those beans are then exported to other Nestlé factories around the world to make localised coffee products. It is thus not difficult to appreciate the close linkages that exist between Nestlé and Colombia's coffee farmers.

Colombia, for its part, is famous for its coffee and the quality of its coffee. *Café de Colombia*, coffee from Colombia, is, as a matter of fact, a brand in itself. For decades, the Coffee Growers Federation has controlled exports, allowing only the highest coffee qualities to be traded abroad. This has contributed to the reputation Colombian coffee enjoys. The crop is at the heart of the country's history, culture, geography, economy and society. It once was Colombia's growth and development engine. This changed drastically in the 1990s, following the dismantling of the international quota system and cyclical price crises. For almost three decades now, structural factors and serendipitous events are jointly hindering the profitability and viability of the Colombian coffee sector.

Falling and volatile coffee prices in international and internal markets, deteriorating productivity rates, diseases such as leaf rust and berry borer destroying coffee trees, accelerated emigration out of rural areas, lack of more intensive and opportune agricultural technical advice, underdeveloped road infrastructure, the mining bonanza, the growth of capital intensive economic activities, fierce international competition and increased climate variability are imperilling the economic viability of Colombia's most representative crop. In the process, the livelihoods of around half a million coffee growers, their families and the communities they belong to are being jeopardised. This sends ripples of uncertainty to the value and supply chains for coffee for Nestlé and other coffee buyers.

This complex scenario suggests that the world's largest coffee buyer may have to worry about its future ability to procure the amount and quality of beans it needs to fulfil an ever escalating demand for coffee products. Clearly, brand and farmers depend on each other to prosper. This is the underpinning premise behind Nestlé's approach to interlocking business and society: 'Creating Shared Value' (CSV). This concept, CSV, was coined and developed by Harvard professors Michael Porter and Mark Kramer (2006, 2011).

Research conducted for this book shows that, for the Swiss corporation, building local clusters, developing products that address social needs and delivering substantial support to supplying farmers is far from being a recent way of doing business. The terminology, on the other hand, is. It was only in 2007 that Nestlé began using CSV as the term of choice to describe the mutual dependence that exists between its operations and society. This is consistent with an earlier study on Nestlé's role in creating milk districts in Moga, India (Biswas et al., 2014). From the time a dairy processing factory was built in Moga in 1961, the company worked with farmers to increase productivity, gain access to quality raw materials, reduce production costs and use local distribution networks.

As such, this book uses Nestlé's conception and implementation of the notion of 'Creating Shared Value' to describe the way in which this important and large player in Colombia's coffee sector has taken substantive measures to secure coffee supplies in the amounts and quality levels it needs. This model also frames the analysis of the company's activities and initiatives carried out in and around the coffee-producing factory in Bugalagrande and the many municipalities from where green coffee is procured.

The company's approach to sustainability in coffee production, processing and retail has been articulated in the integrated Nescafé Plan and Nespresso AAA Sustainable Quality™ frameworks. These two initiatives are thoroughly discussed in individual chapters. After years of implementation, both schemes exemplify how improved day-to-day agricultural practices can translate into higher incomes, more rewarding individual opportunities and improved collective social wellbeing.

Farmers and coffee stakeholders in Colombia recognise the positive contributions Nestlé has made to the national coffee trade. Yet, coffee growers are still a long way from leading prosperous and healthy lives. Ecosystems still need caring for. Structural factors – emigration from coffee regions, ageing farmers, underinvestment, poor infrastructure, low productivity and climate change – are yet to be addressed. Naturally, this is a task the company cannot and should not undertake on its own. As it will be discussed in further sections, whatever progress has been achieved to make the coffee sector more economically, socially and environmentally sustainable has been largely possible due to the establishment of collaborative partnerships with the government, foundations, civil society groups, other companies, consumers and the society at large. Future gains will also be attained in this manner.

Nevertheless, the company remains an indisputably key actor in the country's coffee industry. It is, after all, the largest buyer of Colombian coffee beans, which suggests it is also linked to the ills and fortunes of a large number of farmers. Moreover, the truly global scale of Nestlé's operations offers a powerful example of why the private sector ought to be deliberately and actively brought into the development arena as a key partner. It can bring successful efforts to scale.

Other sustainability-oriented initiatives and certifications have clearly contributed to improving market conditions and living standards for many coffee growers in Colombia and elsewhere in the world. The role played by associations such as Fairtrade, UTZ, Rainforest Alliance and the verification standard 4C is also discussed. This book, nevertheless, centres on how, over time, Nestlé has looked after its interests as a coffee buyer and processor by making coffee production a more profitable and dignified activity for farmers and a more positive activity for the environment. It explores why Nestlé is concerned about societal needs and how these have been included in its operations across the coffee supply and value chain.

This study based its findings on an extensive literature review on the global coffee market, trends, developments, certification and verification achievements and challenges the sector faces at present and for the foreseeable future. It also draws in detail from quantitative and qualitative data gathered during personal and group interviews and correspondence with mainstream coffee producers and also farmers belonging to the Nescafé Plan and the Nespresso AAA Sustainable Quality™ Program in Colombia.

Equally valuable information and insights were obtained from extensive discussions with various staff members at the managerial, professional, administrative and operational levels in regional co-operatives and producers' associations, exporters, local traders, technicians, and Nestlé Colombia Bogota headquarters offices and the coffee-producing Bugalagrande factory. Collected information was processed using data reduction techniques in order to identify crosscutting issues regarding Nestlé's role in coffee production in Colombia.

In October 2013, Nestlé and the Inter-American Development Bank co-hosted the fifth edition of Nestle's CSV Forum, in which participants outlined 'The changing role of business in development'. During the event, the study team had the opportunity to meet with some of the initial interviewees, who provided additional as well as up-to-date information. At that time, the research's initial findings were also presented and discussed. Finally, to test the study's concluding findings and correct any mistakes in the data collected and the way it was interpreted, key actors and two reviewers were asked to give their opinion on the manuscript.

During the research and writing process for this book, it became clear that coffee has become a poster child for sustainability and social justice. For decades now, consumers, NGOs, shareholders and stakeholders have been moving towards closer collaborations to make the sector more sustainable and profitable for farmers. Private buyers and large companies have also joined these efforts. More often than not, they have taken the lead role in setting, enforcing and monitoring a series of product and process sustainability parameters for all actors in the value chain to meet. The rationale behind such actions is clear. If smallholders cannot keep quality and output levels, raise productivity, increase yields, maintain adequate agricultural conditions for crop production and ultimately make a living out of agriculture, businesses will also lose. After all, resource shortages and unstable quality lead to price increases and volatility. No business likes that.

Firms are thinking whether they can make a profit and make the world a better place at the same time. A growing consumer segment is looking to purchase credence goods made sustainably. NGOs and civil organisations are pushing for the trade system to be fairer. Farmers are trying to find partners they can work with to move away from subsistence and into sustained and sustainable profitability. Governments are working to promote development in rural areas. Together, they can all make coffee production a more attractive, profitable, dignified, competitive and professional economic activity. We hope that this case study persuades sceptical and critical readers that social equity, profitability, environmental conservation, consumer satisfaction and business efficiency do not have to be at odds with each other. Nestlé's actions in Colombia are described here as an encouraging example.

Acknowledgments

The highlight of this project has been, without any doubt, the actual visit to Colombia's coffee growing regions, the people who kindly spent time with us, the Nestlé and Nespresso teams who accompanied us, the National Coffee Growers' Federation staff members that guided us through their work, co-operatives and the staff at the Central Mill. The team was fortunate enough to spend countless hours travelling Colombia's roads, visiting coffee farms, talking to farmers and to a wide range of people involved in the coffee sector, and enjoying the breathtaking beauty of the lavish and mesmerising landscape that is home to what we are now convinced has to be some of the world's best coffee. It is to the coffee farmers themselves that we are the most grateful.

The study team would like to express our appreciation to Manuel Andrés, President, Nestlé de Colombia for sharing with us the company's rich and expanding profile of past and present and future activities plans. Equally, we are grateful to the entire staff at the Nestlé Bugalagrande factory, especially Dr Mauricio Trujillo, Factory Manager, and Dr Luis Fernando Victoria, External Affairs Manager. Both management and workers kindly shared with us their opinions and insights about the role the company's industrial presence has had in the municipality and neighbouring areas. They also shared personal anecdotes of the impacts – at the household, individual and intergenerational level – they have witnessed and experienced since they joined Nestlé as employees.

These views were further enriched with candid and insightful discussions with Carlos Alberto Taguado Trochez, Mayor of Bugalagrande, Dr Juan Guillermo Vallejo Ángel, Chairman of the Chamber of Commerce of Tuluá, Juan Manuel Durán Castro, Chairman of the Board of the Foundation for the Integral Development of Valle, and Paula Soto, Director of the Regional Environmental Directorate of Buga. We owe our understanding of Bugalagrande's current peace-building scenario to the team of ladies in charge of the Reconciliation Centre, who showed us the social changes that can be encouraged and promoted when their inexorable commitment to making their communities more peaceful partners with equally minded individuals, institutions and enterprises.

Prior, during and after our visit, we had the opportunity and privilege of working with talented, committed and hardworking professionals and charming, passionate and exceptional individuals. This study would have never been the same had it not been for the relentless support and valuable sector insights both Ricardo Piedrahita and Susana Robledo freely shared with us. Their experience and outlook was thoroughly complemented with thought-provoking and the most enjoyable of conversations with Carlos Rojas Gaitán, Executive Director of Asoexport. Their friendship has been one of the biggest rewards this book has left us with.

Equal thanks to Francisco Bustamante and Valentina Lozano and the entire Cafexport team for their informative and insightful conversations and presentations. With the help of Elisa Botero, and her exceptionally effective logistical arrangements we were at all times able to focus on the study within the overall framework of the country's cultural richness, geographical beauty and culinary delicacies that make Colombia such an exciting place to visit.

We would also like to thank the editor and two reviewers. Their insightful comments were useful to improve the final version of this book.

We should candidly admit that when the Third World Centre for Water Management embarked on this study, each member of the research team experienced different reactions. Whilst safety was initially a concern, we quickly became aware that whatever information on Colombia makes it to the international media it is only a minute portrait of what the country really is experiencing. The field visit turned out to be a stimulating and rich adventure. We would also like to express our appreciation to the security staff that accompanied us at all times. Even though we always felt safe at every place we visited, they cared for us in a discrete but effective and kind manner.

We shall be forever grateful to the numerous coffee farmers we met in Sevilla, Buga, Valle, Manizales, Supía, Riosucio, Jardín and Rionegro. They all received us warmly, and frankly and passionately shared with us their personal stories and those of their families, the perils of the trade, the satisfactions they gain from this activity and their outlook and hopes and concerns for the future.

They shared with us individual, household and community-wide tales of poverty, isolation, hardship, conflict, displacement and violence. This was only a snapshot of the situation to which millions of farmers around the world are no strangers. However, these experiences were overshadowed by the sense of personal perseverance, collective resilience, national remembrance and global recognition that has made Colombia, its coffee and its coffee farmers famous. We are convinced that the history binding Colombia, coffee, coffee farmers, consumers and the other actors shaping this relationship is a tale the world needs to know about.

1   Why Creating Shared Value?

National and international development organisations, agencies and banks; non-government organisations; and even governments have begun to work with third parties to gain access to the financial, human and institutional resources they need to tackle poverty, malnutrition, underdevelopment, increasing inequality, environmental degradation and social upheaval. At once, consumers all around the world are looking at products in a more critical and holistic manner, preferring goods that can simultaneously satisfy their economic, social, environmental and ethical requirements. On their part, firms are seeking to stay profitable and expand their business over the long term. It has been assumed that the agendas followed by all these actors are mutually exclusive and are at odds with each other.

In fact, some believe private sector companies are prospering at the expense of the broader community, greatly disregarding the huge social, environmental and economic problems they cause. Businesses have been effectively disparaged. The narrow conception of capitalism under which many companies operate has curtailed their legitimacy, something especially true for large corporations. Multinationals are accused of manipulating sales prices, tax avoidance, supply chain mismanagement, harmful environmental practices, human rights violations, poor working conditions, murder of union officials, animal rights and welfare abuse, food safety disregard, chemical use, irresponsible marketing, supporting oppressive political regimes, etc. (Chandler, 2008; Sustainability Education Network, 2010; Ethical Consumer, 2013).

Enterprises are being subjected to closer and ever more exigent scrutiny from consumers, suppliers, shareholders, the media, special interest groups and the general public. These different groups want to know how firms build their supply chains, how are benefits and risks distributed amongst shareholders and stakeholders, what are the interactions with the local communities, and how they are becoming more socially responsible and responsive (Bright, 2008; UN Global Compact, 2010, 2013, 2014).

Gradually, a wider range of community issues has begun to gain attention with local and international civil society groups and the media. As public awareness on the impact the private sector can potentially have on public policy community affairs increases, companies are being urged to join other social institutions in improving the living conditions of society at large, especially of the poor and underprivileged (UN Global Compact, 2013, 2014; Falck and Heblich, 2007). The development sector is not staying behind.

In the foreword of the UN Global Compact 2013 Annual Report, United Nations Secretary-General Ban Ki-Moon recognised the work companies have put into 'ingrain universal principles on human rights, labour, environment and anti-corruption in their management and operations'. To 2014, over 2,000 companies using Corporate Social Responsibility, Creating Shared Value, responsible capitalism or sustainable business models had signed up to uphold the standards listed in the United Nations Global Compact. This UN initiative encourages private sector firms and civil organisations from around the world to support good governance principles and adopt environmentally and socially sustainable practices.

In more affluent societies, consumers are looking at products in a more holistic manner than ever before. Quality, price, packaging, the company's reputation, nutrition, wellness and lifestyle portrayal are some of the many factors that come into play when opting to purchase a given product. For many, the demand for goods and services goes well beyond the initial and mere satisfaction of needs and wants. Consumers are often trying to adopt consumption patterns and behaviours that can simultaneously make economically, socially, environmentally and ethically reputable claims.

In developed countries, around 30 per cent of buyers are particularly interested in developing a good understanding of and stand on sustainability issues, the current trade system, rural policies, the workings of the coffee sector, etc. A much smaller number of people, around 5 per cent, of people who claim they are aware of social and environmental issues, rely on those values to make purchasing decisions. This wide 'attitude-action' or 'values-behaviour' gap suggests sustainable consumption is driven by factors other than just moral concerns. For sustainability advocates, most of the members of the middle classes, cost–benefit considerations, socio-economic context, larger consumer trends, marketing campaigns, the media and general consumption behaviours are as important as values and norms (Kennedy et al., 2009; Young et al., 2009 in Kolk, 2012).

The interplay of affluence, consumption behaviour and the growth of a more development-aware socio-economic group has important considerations for sustainability. Each year and until 2030, the World Economic Forum (WEF) has estimated that at least 150 million people will be entering the middle-income bracket (2014). These more affluent and engaged consumers are becoming progressively more concerned with the impacts their consumptive decisions may have on sustainability, the environment and the quality of life of those working in the different steps that ultimately transform raw materials into finished products. They are thus using their purchasing power to push companies to deliver goods and services in a socially responsible and accountable manner, generate positive community dynamics and have neutral environmental impacts. Important as they are in advocating in favour of sustainable agendas, present and future consumers are not the only ones stewarding initiatives to change consumption patterns and make them more sustainable.

Corporate Social Responsibility initiatives can be used as proxies to measure this change in behaviour amongst private sector companies. A 2008 McKinsey & Co. survey found that 90 per cent of the companies engaged stated they had increased their CSR activities from what they were five years ago. More firms began including environmental, social and governance-related issues into business strategies (Bright, 2008). Talks of sustainability and responsibility have gone past boardroom walls to begin permeating corporate goals, missions, visions, statements, press releases and public relations. It can be confidently inferred that since that survey was published, similar activities have grown in number and scope.

Amongst the numerous actual and potential benefits companies can reap by embracing CSR, CSV and the myriad of terms referring to symbiotic business–society relations are increased trust, public reputation and market value. Investment communities are gradually factoring in environmental stewardship, social responsibility and good governance in their calculations of a company's long-term value. Additionally, firms can also expect to enlarge their growth potential from the implementation of 'green' innovations and 'base of the pyramid business ventures' (Guthrie, 2014). In general terms, it seems that acting ethically, responsibly and sustainably raises a company's internal and external profile. It helps build corporate reputation, boosts employee satisfaction and dedication and contributes towards gaining customer trust and loyalty (Ethical Corporation, 2014). It also prevents supplier failure and improves supply chain management. For Nestlé, this is actually one of the main reasons its two largely successful coffee brands have engaged in sustainability-oriented initiatives.

Internal and external pressures alike are persuading firms to implement behavioural changes, including strengthening their supply chains and making their operations economically, socially and environmentally sustainable. Companies naturally want to grow and be continuously profitable over the long term. A difference worth noting is that the activities they carry out and decisions they take in the pursuit of profitability are being increasingly put under regular and punctilious scrutiny by people inside and outside the firm itself. For example, company employees are emerging as corporate agents of change.

Company employees are also consumers, thereby playing dual roles in pushing for sustainability. Many workers are interested in equal measure in consuming goods that are responsibly produced and in being part of a company that manufactures goods and delivers services sustainably (Falck and Heblich, 2007). These roles as workers, consumers and members of society widen the fronts from where sustainability demands are coming. Acting responsibly and sustainably is thus becoming a way for firms to attract and retain human capital.

Additionally, outside actors, especially civil, non-governmental and not-for-profit organisations are becoming part of the global environment in which firms operate (Kytle and Ruggie, 2005). Communities hosting factories, supplying raw materials and providing labour are demanding firms to do good and to consider their demands, needs and be mindful of the challenges they face. All of these new dynamics are shifting power relations and creating new and greater connectivity avenues among stakeholders.

Competitive imperatives are thus undergoing a significant transformation. Companies are gradually listening to, considering and actively addressing social risks, issues, problems and impacts of corporate decisions, sourcing models and factory operations. Building a competitive edge is urging firms to modify their corporate strategies. They are creating blueprint plans for sustainability, sustainable growth and sustainable profitability. Growth models are beginning to incorporate the management of social uncertainties and mitigation of negative externalities. They are also paying closer attention to their supply and value chains.

Firms have increasingly realised that they are closely linked to the destiny of a vast network of supplying farmers, employees, ancillary firms and society at large. It is important to outline the reasons why a company would self-interestedly look after this extended group of stakeholders. Societal and environmental problems can create economic costs in the firm's value chain. Externalities inflict internal costs on the firm.

Firms want to reduce the economic and social risks of producer failure at any stage of the production process, supply chains and marketing. Securing reliable and sustained access to raw materials at competitive rates is crucial to remain profitably in business over the long term (Porter and Kramer, 2006, 2011). Most consumer goods in the world require raw agricultural materials of one sort or the other. These inputs are mostly purveyed from small farmers, working their plots of land at subsistence levels. It is they who are responsible for underpinning local, national and ultimately global supply chains. The absence of supply chains simply means no business in the same way that constrained growth of those procurement steps has negative repercussions for present and future corporate growth potential.

More importantly, the antagonism that places business and people on opposite sides of the development scale is being put into question. Business people, academics, society, stakeholders and shareholders are increasingly thinking about who and what are companies for. Partial and ambitious answers alike have emerged in the form of new business propositions and models that help firms earn their legitimacy. Businesses are enlarging the number of groups they engage with and deepening the way in which they serve them. This process has called the attention of a diverse and growing number of stakeholders and shareholders. It has also spearheaded the emergence of an increasingly vast body of literature that looks at the role private companies are playing in fostering development, new community engagement dynamics, socially focused business models, etc.

Academia, think tanks and business-oriented consulting firms have begun to pay closer attention to the merits, demerits and potential of initiatives such as

Corporate Social Responsibility, Creating Shared Value and Responsible Investment to name just a few. These analyses usually revolve around one key issue: the motivation behind a company building stronger ties with its surrounding community and with actors other than its shareholders.

Procuring the amount and quality of beans it requires to satisfy the growing demand for coffee in the world is proving gradually more difficult. For expansion to be possible and sustainable for many years into the future, coffee producers need to be able to grow sufficient good quality coffee. To do that, they need to sustain a prosperous life at the same time.

When a company supports peasants to be more productive, the benefits are not only for the farming communities or local agricultural systems to reap. Multinational business themselves stand to gain a great deal from investments in physical, human and financial capital incurred to improve practices in the agricultural sector. These actions are in fact an imperative if companies want their supply chains to be robust, reliable, expandable and long lasting. For Nestlé, articulating these needs and demands has taken the form of 'Creating Shared Value', a term coined by Michael Porter and Mark Kramer.

These two scholars defined the term as the 'policies and operating practices that enhance the competitiveness of a company while simultaneously advancing the economic and social conditions in the communities in which it operates' (2011: 66). This 'Big Idea', published in an article titled 'How to reinvent capitalism – and unleash a wave of innovation and growth' in the *Harvard Business Review*, has generated a prolific debate about the validity of this and other models in which societal and corporate businesses intersect, notably Corporate Social Responsibility.

It should be noted that as conceived by Porter and Kramer, and implemented by Nestlé, Creating Shared Value has to be embedded in the company's corporate culture and be treated as integral to corporate strategy. This enlarges the firm's notions of success, urging the private sector to engage in important development debates in order to define and enhance its competitiveness, and simultaneously create economic value and societal benefits within the spheres it operates. This new way of doing business builds upon the notion that social improvement and progress do not thwart corporate growth and profitability, but rather they are mutually reinforcing processes. Moreover, this corporate proposition not only recognises that public priorities, social needs and private goals overlap, but also creates market opportunities in areas otherwise perceived as structural challenges and risks to business.

At its top managerial and leadership level, the company has for some years fully internalised its role as a development stakeholder and partner, firmly cementing its view that development is good for business in the same way that business is good for development. Known as 'Creating Shared Value' (CSV), this business proposition acknowledges that economic growth, social development and environmental protection can only be achieved through collaborative processes that involve the private, public and civil sectors of society (Porter and Kramer, 2006, 2011). It is also about making the private sector grow and prosper, but with a purpose, as the Inter-American Development Bank (IDB) puts it (2013a).

For Nestlé, CSV has become the way it builds and harnesses interdependent relations with farmers, consumers, shareholders and the community at large to grow as a company and build better-off communities. This framework enlarges the intersection of business profitability, social progress and environmental sustainability. Other firms seeking to tighten the relationship they have with society have named their strategies differently: Corporate Social Responsibility, Responsible Capitalism, Sustainable Capitalism, Sustainable Business Development, etc. Many other similar concepts can be found in academic works, corporate mission statements and the media. All these terms seems to have a common genesis. They have emerged from external pressure from costumers and interest groups; internal needs to mitigate supply chain risks and increase employee satisfaction; workers' desire to articulate their roles as employees and community members; genuine corporate efforts to have more amicable relations with society; as well as from personal ideals, individual ethical commitments.

It is beyond the scope of this book to make a categorical statement as to whether CSV is the theoretically more robust model to simultaneously pursue business, social and environmental goals. Other interesting topics not discussed here have to do with the criticisms made to CSV as a concept. Many authors have persuasively engaged in that debate and can substantiate their claims with a robust body of work. Crane and colleagues (2014) have published some of the most vehement responses to the conceptual proposition made by Porter and Kramer (2006, 2011). These scholars see 'Creating Shared Value' as unoriginal and overlooking the tensions between social and economic goals. They find the way the term was conceived and developed by Porter and Kramer as naive about the challenges of business compliance and gives corporations a shallow role to play in society. They see it as a 'reactionary rather than transformational response to the crisis of capitalism' (2014: 131).

The analysis and debate required to reach any conclusion would probably require writing a few volumes on the issue. The vastness and diversity of the corporate world is such that academicians, managers, stakeholders, NGOs and the society at large cannot afford making sweeping generalisations. And yet, categorical statements about the validity, truthfulness and relevance of concepts such as CSR, CSV, responsible capitalism and sustainable business development are all too often found in academic works and widely held narratives. Instead, this case study focuses on the way a large transnational company has gradually amassed a deeper understanding of the links that tie its prosperity to and with society and the environment. More specifically, the arguments presented here are concerned with understanding why and how, through the implementation of its CSV strategy, a large company like Nestlé has become so actively involved in revitalising the coffee sector in Colombia. For example, as notions of corporate success have been enlarged, the company is taking part in rural development processes where new knowledge, capacities and connectivity can create new wealth.

It cannot be stressed enough that CSV is certainly not a panacea for social problems, development constraints, environmental degradation and poverty. What it can do, nonetheless, is to reorient the way a company can do different things and do things differently to reassemble supply chains in a way that is more environmentally sustainable, socially responsible and economically profitable. If implemented with the input from multiple and diverse stakeholders and shareholders, companies can mitigate social conflicts, environmental degradation and disadvantageous economic conditions for the communities they rely on.

Looking into a sustainable and profitable future, Nestlé is seeking to establish a vast community of interests with the different groups it interacts with. These include shareholders, workers and their families, unions, different levels of government, regulators, media, consumers, distributors, suppliers, non-governmental organisations and groups representing the food industry and customers (Nestlé Colombia, 2013b). If all stakeholders recognise and capitalise on the fact that public priorities, social needs and private goals largely intersect, economic value and societal benefits can be simultaneously created.

As a business model, CSV urges the company to develop a comprehensive understanding of its contributions to and impacts on society. It thus requires forging interdependent company–supplier–worker–community–shareholder–customer relationships that can simultaneously lead to improved community welfare and economic success. To successfully do business in Colombia, Nestlé has joined and worked alongside many public, private, social and academic institutions to further advance health, wellness, nutrition and living conditions for consumers and suppliers alike. The company has already established many of the key partnerships it relies on to operate a successful business. Yet, as this study will show in later sections, it will prove to be a challenging task to find partners that share this philosophy and can consequently support its long-term goals.

By adopting CSV as a corporate strategy, Nestlé has had to open up and use all the avenues available to foster intensive dialogues with its multitude of stakeholders and communities of interests. The resulting conversation is leading the company to take ever-greater responsibility for the social, environmental and economic implications of its value chains. This is naturally shifting governance dynamics in local, national, regional and global value chains as well as in local communities.

To redefine its position within society, Nestlé will have to position itself as a proactive, resourceful and innovative development stakeholder and partner. As this book discusses in later sections, its commitment and actions, large and small, are already starting to extend from local initiatives to national schemes and then further to global strategies. This corporate role is helping communities to devise solutions in an increasingly complex and intertwined development panorama. It is bringing a wide array of actors and stakeholders together and enabling and encouraging them to pool their resources and expertise to solve their current and future problems.

Tackling societal challenges has become a way in which Nestlé is transforming its comparative advantage and outperforming its competitors, locally and globally (Pfitzer et al., 2013; Porter et al., n.d.). For instance, given the company's profile and mission, the conception, procurement of raw materials, manufacture and distribution of its products focuses on improving health and nutrition, protecting water sources and the environment and alleviating poverty by fostering rural development. As a result, Nestlé has consolidated its role as a global leader in nutrition, health and wellness, and is furthering its already strong position as the world's largest food and beverages company.

Nestlé is a vast corporation. This study is looking at only one input, in one country, and only one of the factories and processing units the company has in Colombia. The way the company has interacted with its host community, supplying farmers, employees and potential and current consumers has naturally changed since the brand first arrived in the country. For the last few years, Nestlé has referred to this way of relating, interacting and responding to stakeholders and shareholders as Creating Shared Value. As a theoretical model, its workings, mechanisms and tenets are far from being set in stone. Nevertheless, it has gained more prominence as a proposition for Nestlé, and other firms, to tackle challenges faced by both the company and society.

This rejects any suggestions that the company has just discovered the interdependency between social and economic benefits. The concept had not been fully theorised even if the company had long implemented it. Take for example Nestlé's first factory in Moga, India, where the company set up a dairy processing unit back in 1961. From the onset of activities, company and community began working together. There were no special terms given to this relationship. Both sides simply saw they needed each other to prosper and grow profitable (Biswas et al., 2014).

Given the rising prominence of businesses in advancing sustainability and promoting development oftentimes making 'public good type investments', governments are ultimately responsible for planning and implementing welfare- and growth-oriented policies. Public institutions should take the lead in providing a favourable framework and clearly defined objectives that can then rally resources and expertise from all economic and social sectors. By working together, this network of partners can design strategies at different scales to better respond to societal needs and aspirations and achieve mutually agreed development goals.

Nevertheless, since there is still a limited amount of academic research assessing the CSV model, the findings presented in this book seek to contribute to the existing body of literature on this new corporate strategy and its impacts, both actual and potential. They also point to the interest the business community has in getting involved and forging partnerships with a wide range of public, private and civil actors to promote inclusive and sustainable economic growth. The analysis also touches on the factors that have urged the private sector to rethink its relationship with host and sourcing countries and communities, employees, consumers, suppliers, investors, government, civil society and the general public. In this way, the study is aiming at narrowing down the knowledge gaps that still remain regarding the way ordinary private sector firms can become impact enterprises, unleash economic growth, lead to social advancement and promote human development.

Moreover, this research project hopes to encourage other scholars and organisations to thoroughly and systematically look at this rising business proposition so that it can fulfil the promising potential it offers to bring about value for firms, individuals, consumers, communities, suppliers and shareholders at scale. Detailed and comprehensive case studies of companies actually implementing this or similar corporate strategies and the impacts CSV has had on host communities are limited, cover very few value chains and consumer goods and concentrate on the experiences of very specific geographic areas or countries.

Finally, little is known about how other development actors have responded to CSV initiatives; the level of technical, technological, capacity building, managerial and project innovation involved; the partnerships that have been established with governments, universities, NGOs and development agencies; and the way local communities have responded to private sector-led development efforts. These are all emerging and increasingly relevant areas that need to be explored, especially to set realistic expectations regarding the contributions businesses can make to improving overall societal welfare and human wellbeing.

2 The world's beverage of choice

Coffee's popularity translates into well over 2.25 billion cups of coffee being drunk every day worldwide (Brog's Product Development, 2012). Its economic importance arises from the fact that coffee is the second most valuable commodity after oil (Global Exchange, 2011). After water, it is one of the most popular beverages that people consume.

Although currently widely grown on a belt between the Tropic of Cancer and the Tropic of Capricorn with the bulk of global coffee production coming from Latin America, coffee is in fact indigenous to Africa. Arabica (*Coffea arabica*) coffee comes from Ethiopia whilst Robusta (*Coffea canephora*) is believed to have originated along the Atlantic Coast and in the Great Lakes Region. Coffee trees grow best in one of the three main geographical clusters: East Africa and the Arabian Peninsula, Latin America and Southeast Asia and the Pacific. Almost all coffee produced and sold commercially belongs to either the Arabica or Robusta varieties. They vary in taste, growing conditions and price.

Arabicas grow exceptionally well in subtropical and equatorial regions, at altitudes of 600 to 2,000 m and ideal average temperatures between 15 to 24°C. Despite being of higher quality and better taste than Robustas, Arabicas are much more vulnerable to poor soils, diseases, pests, bad handling and climate variations. Beans belonging to this variety contain about 1 per cent caffeine per weight (Coffee Research Institute, 2014).

Robusta, on the other hand, flourishes at lower elevations, between 200 and 800 m above sea level, in an area 10° north and south of the equator, where average temperatures range between 24 to 30°C. These conditions produce coffee beans of sturdier, harsher flavours. Robusta plants are more pest and climate resistant, are less susceptible to rough handling and produce higher yields at lower production costs (Coffee Research Institute, 2014.; ICO, n.d.). It has been commercially on the market since World War II, when, due to its stronger taste and higher caffeine content, it was introduced as lower grade filler in blends (Linton, 2005).

All around the world, Arabicas and Robustas constitute the source of livelihood of some 25 million small farmers and 10 million more workers produce, transport, sell, trade and roast commercially grown coffee in over 80 coffee-producing countries (Panhuysen and Pierrot, 2014; ICO, n.d.). Arabica accounts for 65 per cent of world production with Robusta representing the remaining 35 per cent. Smallholder coffee farmers and their families working plots usually no larger than 5 ha produce over 70 per cent of the total amount of coffee available and traded worldwide.

Amongst all coffee cultivating countries, Brazil is the most successful producer for both Arabica and Robusta varieties. It is also the main exporter of instant coffee. In the last ten years, the buoyancy of the sector has been prompted by the adoption of new mechanisation technologies and novel production techniques. These advancements have triggered a 50 per cent increase in total output, without any increases in cultivated land. The second largest producer is Vietnam. These two countries are responsible for 50 per cent of the global coffee output but the lands they allocate to coffee production add up to only 25 per cent of the world's total land where this crop is cultivated. The third spot periodically changes. Depending on the year, production figures in Indonesia may relegate Colombia's coffee production to fourth place. Colombia, however, is the principal producer of mild washed Arabicas, aroma- and flavour-rich beans of the highest quality.

Colombia's geographic and climatic conditions allow for coffee to be regularly harvested. Even when the crop is evenly available for export all throughout the year, there are two main seasons during which most of the coffee beans are picked: one occurs from September to December, and 'de mitaca', a shorter harvest, takes place between April and June. These two seasons follow two main flowering periods, one from January to March and the second one from July to September. In all other major producer countries, coffee is a seasonal crop. For instance, 86 per cent of Brazil's Arabica beans are harvested from July to December each year and 75 per cent of Robusta obtained from January to June. Around 55 per cent of Vietnam's Robusta harvest is usually available for export between January and June each year (ITC, 2011).

For many other producing countries, these exports make considerable contributions to national income. In smaller procurement markets, coffee is a key source of foreign exchange receipts, gross domestic product and fiscal contributions in the form of export taxes. During the 2005–2010 period, eight countries around the world relied on coffee exports for more than 10 per cent of their total export receipts (ITC, 2011). According to the International Coffee Organisation (n.d.), Timor Leste's share of coffee exports as a fraction of total exports, by value, reached 70 per cent during 2005–2010; it was followed by Burundi (62 per cent), Ethiopia (34 per cent), Rwanda (28 per cent), Honduras (21 per cent), Nicaragua (18 per cent), Uganda (17 per cent) and Guatemala (13 per cent). Coffee is clearly a crucial commodity for these and other countries. It is also key for global trade.

For the world's top coffee producers, the share this crop represents in their balance of trade is relatively low, even if considerable. The figure reached 3 per cent for both Brazil and Vietnam (ICO, n.d.). In Colombia, coffee exports brought 3.2 per cent of total export income in both 2012 and 2013. Just three years prior, a rise in international coffee prices pushed this share upwards. Coffee exports represented 7.4 per cent in 2009 and 5.1 per cent in 2010 and 2011 (DANE, 2013). Much harder to quantify is the historical, social, cultural and environmental importance the crop has for these and other producing countries.

In 2012, the world consumed close to 142 million bags of coffee. As an international standard, each bag of coffee is 60 kg. Close to 80 per cent of it was traded internationally. The export value of this flow amounted to USD 33.4 billion, fetching three times as much, some USD 100 billion, in retail sales. In 2009 alone, the soluble coffee market had a value of more than CHF 10.6 billion (ITC, 2011). This profitable market did not exist 75 years ago. It was in a way created by Nestlé. The largest direct buyer of green coffee in the industry has at least twice revolutionised how and where people drink coffee. The company introduced the world's first instant coffee, Nescafé®, which is now the fifth most valuable food and drink brand globally and the world's best selling coffee brand by a factor of five (Nestlé SA, 2014d). In the premium market segment, Nespresso has become a reference for the portioned coffee capsule system.

The innovation path that began in the 1930s when instant coffee was first launched has not ended. New coffee products have been developed and established ones renovated. Consumers have gained access to breakthrough coffee systems that recreate coffee-shop experiences at home. This transformation has extended to production systems and supply chains. As will be discussed in later chapters, the most important change in the coffee value chain has taken place in how and which type of coffee is sourced, from whom and to what have been the impacts in society, rural economies and the environment.

Redefining the coffee sector

Twice in its history, Nestlé has radically transformed the global coffee sector. As early as coffee came to be known to a wider group of consumers in every corner of the world, the quest began to find a way of preserving it. Retaining its taste and aroma posed a challenge of similar or even greater magnitude. In 1901, a Japanese chemist invented instant coffee but the product met little marketing success. Elsewhere in the world, and in the early 1900s, food scientists had patented liquid and powdered coffee for the mass market. Despite the novelty of the new product, the commercialisation and popularity of such drinks remained very limited. The taste was simply unappealing.

A few decades later, in 1930, the Brazilian Coffee Institute contacted Nestlé to develop granulated coffee, soluble in hot water and that retained its flavour. The country was stocking a large coffee surplus that ran the risk of going to waste. Through a collaborative approach with the Brazilian Government, and seven years of research, a group of Nestlé scientists, led by chemist Dr Max

Morgenthaler, developed Nescafé®. A portmanteau of the words 'Nestlé' and 'café', the breakthrough product was announced and introduced on 1 April 1938 at Nestlé's headquarters in Vevey. This new process permitted large-scale industrialised production and transformed the industry dramatically by avoiding crop waste. This technology was first applied at the Orbe factory in Switzerland and then at the Hayes factory, in West London, UK (History of Business, 2008).

By 1940, Nescafé® was already available in 30 countries. Shortly after soluble coffee was first exported to France, the United Kingdom and the United States the new product swiftly became a staple beverage for consumers around the world. During World War II, Switzerland, the United Kingdom and the United States alone consumed more than three-quarters of all Nescafé® coffee produced globally. The lion's share was destined for American troops, as Nescafé® was an integral part of their field rations. This contributed to popularise the concept of coffee as a drink and made American forces 'brand ambassadors' in Europe (Nescafé, 2013). Throughout the war, the Nestlé plants in the United States reserved their entire annual production of one million yearly cases for the army (Razeghi, 2008; Monson, 1991; The Blade, 1980).

Since World War II, and propelled by further innovation, coffee's popularity as a drink has grown. In 1952, the St Menet factory in France developed a type of instant coffee that did not need additional carbohydrates. In 1960, Nescafé® was re-launched in the now characteristic glass jar. Five years later, in 1965, the Gold Blend variant introduced freeze-dried soluble coffee and, in 1967, coffee granules replaced coffee powder. By 1994, Nestlé had developed the full aroma process, greatly improving the quality of its instant coffee and thus further enhancing its global attractiveness.

Constant improvements and new ideas have firmly positioned Nescafé® as the world's leading coffee brand. This ranking is strongly backstopped by the brand's commitment to quality and taste. Most Nescafé® blends use Arabica beans, at least partially. Nescafé® GOLD is a blend of Arabica and Robusta beans; Nescafé® CAP COLOMBIE exclusively uses Colombian Arabica beans; Nescafé® ALTA RICA blends Arabica beans sourced from Latin American countries only; and Nescafé Espresso blends Arabica coffee beans from various regions of the world (Nescafé, 2013).

In the coffee industry, Nestlé has built a strong reputation around its capacity to respond to the needs and wants for individualisation that customers look for in the market products they purchase. With Nespresso, one of Nestlé's most profitable commercial brands and one of the most successful business model innovations of recent years, the company has reached a unique market positioning. This success is the result of the convergence of customers' requirements for customisation, the ready availability of a wide range of high quality coffee varieties and flavours, and convenience (Matzler et al., 2013). Instead of limiting the number of coffee tastes a consumer could enjoy at home at one given point in time with a 250–500 g pack of coffee, Nespresso has expanded the range of individual experiences for individual coffee drinkers. New formats have now been developed to offer this portioned coffee capsule system to customer- and service-oriented high-end businesses such as hotels, restaurants and offices.

Developed in the 1970s for the business market, Nespresso SA initially sold two machine types and four coffee varieties. It was then commercially launched in 1986, but took a decade for Nespresso to reach break-even point and revolutionise a sector that seemed to be entirely dominated by instant coffee. This created a new category altogether: the capsule coffee market. Consumers can create a cup of coffee that is up to the standards of a skilled barista whilst being able to conveniently individualise their espressos according to their personal preferences. Coffee drinkers can choose from a range of 22 different Grand Crus available at all times and several limited editions released every year.

Grand Crus indicate exceptional quality throughout the coffee production process. Whether from a single origin or blends of various varieties and places, beans are sourced from regions with ideal soils and climate, controlled growing conditions, quality- and sustainability-oriented farming practices, and compliance with specific discriminatory tasting and quality controls (Nestlé Nespresso SA, 2014a). Nespresso's Grand Crus are classified into six different groups according to the intensity of the coffee used: espresso, intenso, pure origin, decaffeinato, lungo and variations.

Espresso Grand Crus include Livanto (with Arabicas from Colombia and Costa Rica), Capriccio (a blend of Brazilian Arabicas and a touch of Robusta), Volluto (100 per cent Colombian and Brazilian Arabicas sourced from the Nespresso AAA Sustainable Quality™ Program) and Cosi (a combination of Arabicas from East Africa and Central and South America). The multi-origin intenso varieties are Kazaar (a blend of two Robustas from Brazil and Guatemala and a washed Arabica from South America), Dharkan (a blend of Arabicas from Latin America and Asia), Ristretto (a blend of Arabicas from Colombia, Brazil and East Africa and Robusta), Arpeggio (a selection of Arabicas from South and Central America) and Roma (with Brazilian Arabicas and Robustas and Central American Arabicas).

Pure origin Crus include Indriya from India (blending Arabicas and Robustas from the sub-continent's southern region), Rosabaya de Colombia, Dulsao do Brasil (with coffee sourced from Arabica plantations from southern Brazil) and Bukeela ka Ethiopia (composed of two different Arabicas from the western part of the country). Decaffeinato coffees include Decaffeinato Intenso (a selection of South American Arabicas), Decaffeinato (a blend of South American Arabicas and Robusta) and Decaffeinato Lungo (a Grand Cru combining Brazilian and Colombian Arabicas and a touch of Robusta). The three lungo varieties are: Fortissio (made from Arabicas from Central and South America and a hint of Robusta), Vivalto (which uses South American and East African Arabicas) and Linizio (a blend of Colombian and Brazilian Arabicas). There are also three Livanto variations: Caramelito, Vanilio and Ciocattino, which became permanent Grand Crus in 2013 (Nestlé Nespresso SA, 2014a).

Nevertheless, the unique espresso experience Nespresso promises to deliver does not depend only on the quality of coffee used to make the capsules. The machine itself plays a crucial role. It controls water temperature and pressure to the constant levels required to brew a good quality espresso. Manufacturing these appliances, however, is a far-flung activity from Nespresso's areas of expertise and comparative advantage. It proved to be much more efficient for the brand to partner up with external designers and engineers to create an extraction system that could reveal all the flavour notes of coffee (Matzler et al., 2013).

This exercise took over ten years and the filing of a large portfolio of patents (Matzler et al., 2013). The result has been the fabrication of elegantly designed, high quality coffee machines sold at an economic price through licensing retailers. Ultimately, the Nespresso extraction system has culminated in the creation of an intuitive and simple coffee preparation process. Capsules are easily inserted and automatically perforated, water pumped optimally and homogeneously at a constant pressure of 19 bars, and used capsules ejected into a collection container (Nestlé Nespresso SA, 2014a).

Appliances are manufactured and distributed through licensing agreements with Nespresso (Matzler et al., 2013). Machine producers DeLonghi, Miele and Krups have been working closely together with Nespresso to fulfil the brand's commitments to environmental sustainability and the reduction of machine energy consumption. Energy efficiency has contributed to cuts in the carbon footprint of a cup of Nespresso coffee. Carbon emissions from the moment coffee is produced, transported, packaged into capsules, distributed to boutiques and brewed in the Nespresso machines have been slashed by 20 per cent. With time, new manufacturing companies have joined the brand's efforts to manufacture more efficient, more stylish and user-friendlier machines.

For example, the VerTech™ Network for sustainable technology research and development project brought together machine suppliers, engineers and sustainable technology experts to design energy-efficient machines. The objective is to improve energy consumption and reduce carbon emissions. Similarly, the CitiZ range received an award for designing appliances with lower energy consumption. The introduction of eco-timers and stand-by switches help to reduce the equipment's carbon footprint. It is important to keep in consideration that Nespresso does not profit directly from the coffee machines. At the core of the brand's success is the coffee used to manufacture the capsules and the new avenues it has developed to reach actual and potential consumers.

The brand's value chain focuses on coffee retailing in previously uncharted ways. Leaving food retail distribution aside, Nespresso introduced the concept of Internet and telephone sales, specialised stores, appealing coffee bars and exclusive retail boutiques. This marketing strategy has consolidated the brand's image of exclusivity and has established a direct link with Nespresso customers via the Nespresso Club (Furrer, 2011). Customers automatically join this Club upon placing their first single serving coffee capsule order.

As a luxury product, Nespresso coffee feeds consumers' lifestyle ambitions by being firmly positioned at the very top of the quality and upmarket coffee pyramid. To 2014, actor George Clooney has been the image in the brand's aspirational 'What Else?' advertisements. The brand initially targeted European markets with a strong tradition of espresso consumption but its popularity has significantly expanded to other regions. Even in the United States, Asia and Latin America, places that have been more generally considered as 'white cup' markets as coffee is drunk with milk, Nespresso's presence has grown considerably.

Worldwide, the brand has grown at an average 30 per cent since 2000, reaching USD 3.8 billion in global sales in 2010 (Nestlé Nespresso SA, 2011c). This expansion is also reflected in the number of people who work for Nespresso, the number of countries where the brand is present and the number of boutiques operating worldwide. In 2000, Nespresso had 330 employees. In 2014, some 14 years later, the team reached 9,500 people. At the end of 2013, Nespresso systems, capsules and coffee-related accessories by renowned designers could be acquired in 327 boutiques in 52 countries. As an ultra premium brand, it has also sponsored major international events such as the Cannes Film Festival and the Americas Cup (Nestlé Nespresso SA, 2014f).

Targeting a different market segment, Nescafé® launched Dolce Gusto® in 2006. This machine system allows consumers to tailor a wide range of coffee-based beverages at home. This off-the-shelf, single-cup capsule system is manufactured by DeLonghi and allows consumers to prepare and personalise a range of 22 recipes of both hot and chilled drinks, ranging from black and milk-based coffee and chocolate drinks, as well as different kinds of teas (Nescafé Dolce Gusto, 2013). Unlike Nespresso, the machine uses capsules that are widely available in supermarkets and other stores. Allowing consumers to explore the diversity of coffee types, Nescafé® Dolce Gusto® uses both roast and ground Arabica and Robusta coffee for its apportioned blends. An important share of the Arabica beans come from Colombia, which makes Dolce Gusto® yet another important buyer for Colombian coffee (Nescafé Dolce Gusto, 2013).

Together, the success Nespresso has had in the upscale market and Dolce Gusto® in the midscale segment have encouraged the market entry of a large number of competitors. What is more, Nespresso pioneered the coffee capsule technology and can lay claim to the popularity this type of system has recently gained. The global market was estimated at USD 10.8 billion in early 2014 (Jolly, 2014). In Western Europe alone, and over the course of only four years (from 2009 to 2013), this relatively new market for pre-apportioned, single-use coffee capsules more than doubled to €4 billion (Daneshkhu, 2013).

Naturally, with such robust rates of market growth, the emergence of competing brands chasing after this profitable and expanding market has not waited. Especially targeting Nespresso consumers, there are currently some 45 competitive systems and 90 brands offering compatible capsules (Nestlé Nespresso SA, 2014f). These generic pods are sold at lower prices and in supermarkets.

Competition clearly challenges Nestlé's dominance in the coffee sector though, more importantly, it draws definitive differences with the overall product experience rival brands offer. Nespresso places a strong focus on quality, innovation and direct-to-consumer service. For consumers in the high retail end seeking exceptional cup profiles, the type of bean quality the brand offers clearly represents a competitive advantage. These quality levels are maintained and assured even when coffee is procured from hundreds of thousands of individual farms from every tropical corner of the world.

Despite being the largest coffee buyer in the world, or perhaps because of it, Nestlé does not own any commercial coffee farms nor does it resort to contract farming. Instead, it sources its coffee beans directly from coffee producers and the associations and co-operatives they form. This community of farmers is built and strengthened through extensive and intensive technical assistance programmes and direct procurement operations, called Farmer Connect. The firm's actual and potential impact on rural development is thus considerable (Nescafé, 2013).

Across the supply and value chain, producers, trading partners and corporations are creating and adding higher value to coffee products. This is in part a reaction to the value consumers attach to the process and social and environmental conditions under which coffee is grown and commercialised. In this way, the demand for speciality and certified coffees, as well as the growth of the portioned or single-serving coffee sector, has intensified. With it, so has the corporate need to secure stable and high quality bean supplies. Sustainability has come to define competitiveness in the coffee sector.

Nespresso and Nescafé® have responded to these market opportunities and sourcing challenges in a manner that identifies social needs and recognises the brand's interdependent relations with farmers, consumers, shareholders and the community at large. This business approach, pioneered by its parent company Nestlé, seeks to create shared value (CSV) all along the supply chain and shareholder and stakeholder pyramid. The company fully recognises that its successful coffee brands depend on thriving farming communities to secure good quality beans. As a result, the company has put in place a coffee-specific comprehensive agenda, the Nescafé Plan and Nespresso's integrated CSV AAA Sustainable Quality™ Program to optimise the company's global coffee supply chain.

Launched in August 2010, and for the next decade, the Nescafé Plan will invest a total of CHF 500 million in coffee projects all around the world, CHF 350 million for Nescafé® and CHF 150 million supporting the Nespresso Ecolaboration™ sustainability platform. The two different initiatives are changing farming practices and lives by identifying and then promoting sustainable production and processing leading to higher productivity, higher incomes and better living conditions (CRECE, 2011). The impacts they are having on Colombian farmers and their communities are discussed in subsequent sections.

3 The shared C in coffee and Colombia

Complexity

Despite the fact that there are over 80 countries in the world producing coffee, for many coffee drinkers around the world, to say coffee is to say Colombia. The country is well known for the aroma and acidity of its mild washed Arabica coffee beans. For Colombians, coffee holds a special place in the cultural, social, economic, political and environmental narrative of the country.

After almost two centuries of using coffee as a motor for social, economic and even political growth, power and change, Colombia currently is the world's third largest producer, after Brazil and Vietnam. Its ranking is closely followed by Indonesia, with whom it changed ranking positions in early 2014 following heavy floods and the resulting loss of output in that country. Colombia, however, is by far the world's principal producer of washed Arabica beans.

Colombia's history with coffee is too vast and intricate to be discussed in detail here. The work of scholars like Bergquist (1986), Griffin (1968), Palacios (1980, 2002, 2006) and Roseberry et al. (1995) document the sector's development throughout the country's contemporary history and the effects it has had on its societal, economic and political development. The so-called 'coffee crisis' in the 1990s radically reshaped the international coffee landscape and the structural conditions faced by many coffee growers in virtually all coffee-growing regions around the world. Authors like Lozano (2010) offer an understandable compilation and literature review of a wide array of studies looking at the socio-economic and technological changes the sector has undergone over the course of the last two decades.

There are several versions narrating the arrival of coffee to Colombia, although most authors seem to attribute its introduction to the Jesuits, around 1730. Before the coffee era, Colombia was still not a nation, it did not have a national monetary system and people depended on large landowners. Coffee came to transform, for good, Colombia's economy, geography and society. Shortly after the crop was introduced in the country in the eighteenth century, internal colonisation spread its cultivation throughout the central regions (National Coffee Park, n.d.).

Local geographical conditions facilitated coffee's introduction into the productive and natural landscape. The process almost took place effortlessly in the country's mountainous and hilly terrain. Mild Arabicas are more fragile varieties and grow best in mountainous ecosystems where coffee plants are nurtured by low pH soils, light shade, even precipitation throughout the year and temperate climates. In Colombia, coffee is grown in soils derived from volcanic ashes, where altitudes average 1,200 metres above sea level and annual precipitation ranges between 1,800 and 2,500 litres per square metre (Perdomo and Mendieta, 2007).

Moreover, and in conjunction with the 95 per cent of farms that have an average size of less than 2 ha, the country's physical characteristics have also given Colombian coffee a notable, albeit labour intensive and demanding feature: beans are harvested by hand (Consejo Empresarial Colombiano para el Desarrollo Sostenible – CECODES, 2012). In the absence of technological devices to mechanise the harvesting process, each cherry is hand picked. Far from amounting to an absolute comparative disadvantage, this trait has turned into a notable feature and an added value element of Colombian coffee production as manual bean selection implies better quality (Roldán-Pérez et al., 2009). The combined result of hand harvesting and the washing process ensure the unique and distinctive aroma, body and taste of Colombian coffees.

The landscape's ecologic conditions were propitious to successfully produce coffee in small, family-owned farms. This created a rural middle class that demanded industrial products on a large scale. More widespread land ownership also made coffee producing regions more democratic and with narrower social disparities (Urrutia, 1979 in Posada and Pontón, 2000). As international prices increased, coffee growers began obtaining higher incomes. Growing affluence soon established them as one of the most influential groups in the country, driving national aggregate demand for goods and services and supporting the development of other economic sectors in Colombia (Cárdenas and Junguito, 2009; Silva, 2004). For a good part of the twentieth century, coffee was the sole agricultural product stimulating economic growth in both rural and urban areas (Bejarano, 1987).

Moreover, whilst larger-scale coffee production started around the 1850s, its impacts could be felt on the global markets by the early 1900s. Colombia's share of global coffee exports reached around 5 per cent. Prior to 1870, coffee-cultivated areas rose slowly even if coffee holdings had begun to prosper considerably, especially in the departments of Cundinamarca and Santander, where 80 per cent of the beans were being produced. In the last decades of the nineteenth century, coffee prices rose and production expanded to the western departments of Antioquia, Caldas and Valle to satisfy export demand, mostly coming from the United States (Urrutia, 1979 in Posada and Pontón, 2000).

Already in 1895–1898, coffee represented two-thirds of Colombian exports (Ocampo, 1979 in Posada and Pontón, 2000). In only 18 years, from 1880 to 1898, the number of bags sent abroad grew fivefold, from 100,000 to 500,000 (Ospina, 1974 in Posada and Pontón, 2000). From 1900 onwards, coffee became an important agricultural Colombian export and an important source of employment for surplus labour in the countryside. Between 1965 and 1995, the country contributed, on average, 13.5 per cent of the world's coffee production. Over that period, coffee exports earned nearly 80 per cent of the country's total foreign currency receipts (Cano et al., 2012: 7). This figure, nevertheless, has moved within a rather wide percentage range and drops considerably when taking into consideration a wider timeframe. For instance, from 1851 to 2010, Colombia's coffee exports averaged 8.94 per cent of the global total, a share that peaked to 25 per cent in 1944 but fell sharply to 19.2 per cent three years later. For a detailed progression of the global share of Colombian coffee exports over the last 150 years see Figure 3.1.

By the late 1930s, Colombian coffee exports accounted for nearly 25 per cent of the global total. This led to multiplier economic effects and stimulated the growth of other domestic industries such as the financial, trade, transport and services sectors (Pizano, 2011). Twenty years later, in 1952, coffee was responsible for nearly 11 per cent of the national gross domestic product (GDP) and over 29 per cent of agricultural GDP. In a domestic economy conditioned by low levels of physical and human capital, coffee exports rose to represent 70 per cent of all exports for two decades, during the 1950s and 1960s, and on average, 61 per cent between 1908 and 1989. Accordingly, Colombia's economic cycle was related to the price of coffee and thus the coffee bonanzas in 1954–1955, 1975–1978 and 1986 led to a few years of robust growth (Cárdenas and Junguito, 2009).

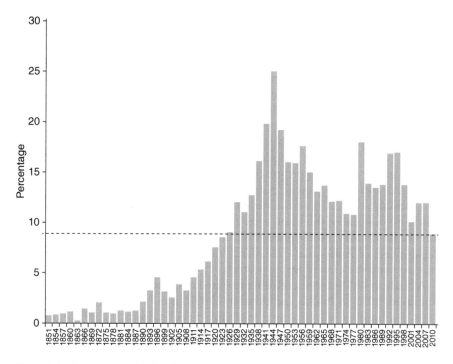

Figure 3.1 Global share of Colombian coffee exports, 1851–2010
Source: Lisboa-Bacha et al. (1993)

Due to internal and external factors, the once very tight relationship between coffee and national income has nevertheless changed throughout the decades. Domestically, even if coffee has remained a key primary product and an important employment generator, it has slowly lost its once undisputed economic and policy influencing muscle. The economic profile of the country has moved away from agriculture. Priority has been awarded to industrial and mining activities. Moreover, the tertiary service sector, albeit informal, has not ceased to thrive. The steep changes in contributions of agriculture and coffee to the Colombian GDP from 1950 to 2009 can be clearly seen in Figure 3.2.

The significant economic role that coffee once played in Colombia's productive activities has steadily declined since the end of the International Coffee Agreement (ICA), the quota system and the collapse of world coffee prices in 1989. In place from 1962 to 1989, the International Coffee Organisation (ICO) quota system regulated the amount of coffee supplied to the international market, withdrawing excess coffee production and controlling prices. Supported by the United Nations, the ICO was established in London in 1963 as an intergovernmental initiative to regulate an international market that had witnessed sharp price volatility in the 1950s and beginning of the 1960s and was thus creating havoc in developing country producers (ITC, 2010; Roldán-Pérez et al., 2009).

Since 1960, Colombia has developed a country brand and has consistently promoted and differentiated Colombian coffee as the best in the world. Fifty years later, it embarked on the '*Cafés Especiales*' strategy to penetrate a new market segment focused on the regional origin of coffee beans. The strategy sought to generate greater revenue for farmers by adding more value at the origin (Reina et al., 2008 in Rueda and Lambin, 2013). Only in the very last

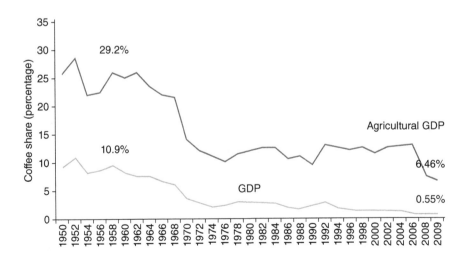

Figure 3.2 Contributions of agriculture and coffee to Colombian GDP, 1950–2009

Source: 1950–1974: Junguito and Pizano (1991); 1975–2009: DANE (2013)

few years have conditions for farmers improved. Domestic productivity levels have begun to rise and with them so has output. In the international market, the international price of coffee has partially recovered as overall world output has experienced ups and downs caused by rust infestations in many producing countries. History, however, should serve as a reminder that this more optimistic scenario is usually short lived and cyclical.

Reviving the coffee sector in Colombia is not solely based on economic or social considerations. Without doubt, the sector is motivated to regain the terrain in the international Arabica coffee market it has lost and other competing countries have seized. Domestically, however, coffee has long ceased to be at the epicentre of the country's development. It no longer is the main national product contributing to GDP in both agricultural and non-agricultural productive activities, nor the main generator of foreign exchange or the most exported commodity it once was. However, unless structural and systemic factors limiting the profitability of agriculture and the rural sector as a whole are not addressed, the Colombian coffee sector, and around half a million coffee growers, their families and the communities they belong to, will still face a precarious situation.

Despite this seemingly adverse scenario, Colombians rightly still think of coffee with a capital C. In the country's popular and historical narrative, coffee is the start and result of multifaceted social, economic, political, technological, migratory, environmental, cultural and organisational forces. At times these factors work in synch with each other. At others, they pull in divergent directions. Coffee stands for rural capital, culture, capacity building, class, community, cohesiveness, coexistence, consistency, challenges, connectivity, communication, consumption, change, continuity and competence. In one word: complexity.

Price volatility

Consumers interested in cup quality unequivocally prefer Arabica coffee, which comes from tastier species but is grown on trees highly susceptible to diseases and changes in soil composition. Conversely, Robusta is more resistant to phytosanitary threats and has higher yields per tree than Arabica but it is of lower quality. Its bolder and stronger taste, as well as higher caffeine contents, means Robusta beans have been well suited for use in the production of instant and flavoured coffees (Panhuysen and Pierrot, 2014). Arabica is produced in smaller volumes than Robusta, on considerably smaller farms and at much higher altitudes. Arabica plots are usually less than 1 ha. In fact, only about 5 per cent of coffee production in the world is of high quality Arabica.

The markets for Arabica and Robusta are highly dependent on each other. Even when the two coffee varieties compete in different segments, they are often substituted and exchanged in blends catering to the mainstream market, this despite considerable differentials in commercialised volumes and prices paid. They are also traded on different stock exchanges. Arabica beans are traded on the New York Stock Exchange, with prices historically moving in a range of 150 to 200 cts/lb. Robusta attracts lower prices, between 80 and 100 cts/lb, and is traded on the London International Financial Futures and Options Exchange (Nestlé Nespresso SA, 2012b, 2013a).

To calculate the international price of Colombian coffee a series of elements need to be factored in. On a daily basis, the Coffee Growers' Federation (Federación Nacional de Cafeteros, FNC), publishes the price at which coffee is trading. This rate is based on the supply and demand for washed Arabicas as captured by the New York Coffee Exchange settlement price. Denominated in ex-dock, American dollars per pound (coffee physically placed in dock warehouses in New York), this price is used as a reference for domestic coffee prices paid in Colombia. An *excelso* coffee premium is then added to this base price as a pecuniary recognition of the superior quality that characterises Colombian coffee (FNC, 2011b). Mild Arabicas have historically sold at a premium above Robusta and normal Arabicas.

Prices in American dollars are subsequently exchanged into Colombian pesos using the last negotiated exchange rate of the day. Lastly, for the price of *excelso* coffee to be converted into ex-dock terms (as placed in storage at buying countries), the costs and storage expenses of placing parchment coffee at the closest Almacafé warehouse are then discounted (FNC, 2011b). Almacafé belongs to the FNC and is the body in charge of operating coffee storage facilities in Colombia. Price variations between different Almacafé storage facilities are due to the distance separating them from the seaports.

Moreover, Arabica price premiums over Robusta have narrowed sharply in recent years. In July 2013, price differentials between the two varieties reached a four-year low of 11.35 cts/lb. Colombian milds, which sell at a premium above the baseline prices paid for regular Arabica, have also seen that price differential collapse. Between 1988 and 2003, Brazilian and Vietnamese coffee exports drove down international coffee prices to a historical low that was 73 per cent below the previous levels. This sharp plunge and historical variations in the international price of coffee can be appreciated in Figure 3.3.

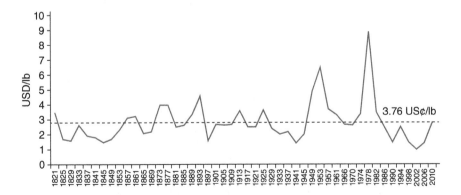

Figure 3.3 Price of Colombian coffee in constant 2010 US dollars
Source: FNC (2013b); Ocampo (1990)

Over the last 20 years, coffee growers have been severely affected by high short-term volatility, deteriorating terms of trade and decreasing real prices. The situation became ever more precarious due to the absence of technological breakthroughs. In the last two decades, no major technological changes have been successfully introduced to increase productivity. Moreover, profitability has also been eroded by the sharp rise in the cost of fertilisers and labour. Even in oil and gas rich Colombia, the prices of chemical inputs represent up to one-third of production costs.

Declining profit margins translate into a series of farm-level adjustments. For instance, reinvestment rates are downwardly revised, a decrease that hurts productivity and yields. Against this backdrop, and exacerbated by growing demand for speciality coffee, sourcing high quality beans becomes a gradually more complex and difficult process. This scenario also suggests that the economic situation of Colombia's domestic economic cycle for coffee is heavily dependent on external economic dynamics on which farmers have limited or no control.

Ageing factors of production

When coffee was first introduced in the western parts of the country, it was successfully grown on family-owned farms. As previously discussed, coffee could be produced on small estates, which helped address social disparities. The more people gained access to land and grew coffee, the more entire communities were able to enjoy better social and economic conditions (Urrutia, 1979 in Posada & Pontón, 2000). With time, the previously generous benefits extracted from growing coffee fell dramatically. Already smallholdings were further divided amongst family members. Access to land, even if only to very small plots, and changing socio-economic conditions were once factors of success in coffee production. They now seriously affect the sector's profitability.

In Colombia, 89 per cent of coffee growers own less than 3 hectares of land. Out of this, an average of 2 to 2.5 ha are used to grow coffee. Plots are traditionally divided and used to grow different crops, coffee being only one of them. Farmers usually complement their income and dietary needs by growing cash crops, primarily fruits. According to the FNC, 60.59 per cent of all farmers grow coffee in less than 1 ha of land, 28.36 per cent do so in 1 to 3 ha and 5.79 per cent in 3 to 5 ha. A small minority of 4.76 per cent cultivates coffee beans in what are considered to be 'medium to large' landholdings even when estates range between 5.1 and 20 ha. Only a very small group, less than 0.5 per cent of all farmers grow coffee in plots larger than 20 ha (Pizano, 2011).

Figure 3.4 shows the distribution of land in coffee growing areas. This depiction clearly illustrates the extent to which once larger plots have been broken into small estates. The vast majority of farmers and their families depend on what these small farms can produce to survive. Even if agricultural productivity were at its maximum possible levels, holdings are so small that it would be disingenuous to expect families of five members or more to sustain an adequate standard of living. Poverty in rural communities is closely linked to access to land, productivity and the international price of coffee.

Such small properties reduce the scope to reap benefits of economies of scales, namely incurring into lower inputs cost to produce higher outputs. Micro farms have also been found to be technically less efficient. In 2007, the Centre of Economic Development Studies (CEDE), University of Los Andres, conducted a Data Envelopment Analysis (DEA) study in the coffee growing regions of Caldas, Quindío and Risaralda. Researchers found that, in general, an average coffee grower was only 42 per cent technically efficient. This percentage fell to 36 per cent for small farmers. This ratio reached 51 per cent among medium-size producers and 60 per cent for large ones. These findings unveil the existence of a large window of opportunity available to farmers in order to improve the way they manage the use of the main inputs used for coffee production, namely labour, chemicals and machinery (Perdomo and Mendieta, 2007).

Other studies have looked at different levels of technical efficiency across coffee producing countries. Perdomo and Mendieta compared similar groups of coffee growers in Colombia and Vietnam. The costs of production per pound of coffee in a Colombian farm were USD 0.57 for large coffee farmers, USD 0.60 cents for medium-size producers and USD 0.61 for smaller farmers. Controlling for differences in types of coffee varieties grown, terrain and other conditioning factors, the technical efficiency index was substantially higher in Vietnam. These results indicate that the Colombian farmers have the potential to increase yields by optimising the levels of inputs used (Perdomo and Mendieta, 2007).

These findings, however, have not convinced all scholars. Lozano (2007) studied the relationship between farm size and productivity per hectare. For

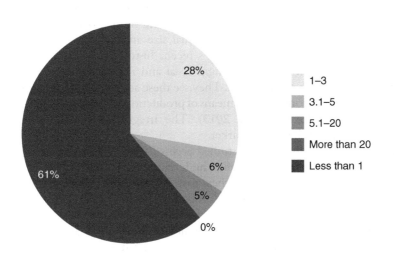

Figure 3.4 Size distributions of coffee land holdings in ha
Source: Pizano (2011)

small producers, mean output tends to be low and marginal productivity below the market wage. The author used data from the survey 'Analysis of Labour Market Coffee and Access to Credit for small coffee producers in Colombia' the FNC had put together in 2006. He also drew from Agricultural Household Models to confirm the existence of an inverse relationship between productivity per hectare and the size of the plantation for producers working in plots 5 ha and smaller. The author found that the smaller the plot, the higher the total production and productivity. Although important, these are not the only factors influencing production levels. Household conformation and characteristics play a determining role.

Farm yields and household income depend on whether women or men are the household heads. Lozano's study also found that production was 45 per cent lower for women-headed households. Lower output levels were also associated with the number of family members working on the farm (Lozano, 2007). Marginal returns to labour diminished steeply due to the minute size of an average coffee plantation. This marked relationship was largely associated with the lack of employment opportunities in rural areas. A lack of work alternatives encourages households to intensify the use of family labour. The close interdependence that interweaves consumption and production decisions suggests households are able to produce their means to their subsistence even when the in-farm generated income is lower than what they could obtain if they sold their labour outside their farms.

Plot size and low technical efficiency are likely to become even more important issues in the coming years depending on how the demographic dynamics of inherited land tenure, land prices and migration interact with each other. If retired farmers further divide their lands to pass them on to their children, already subsistence-size plots are likely to become even smaller. This could be a serious constraint to Colombia's overall coffee production. It should also be noted that farmers are not always trustful of the government and thus land titles do not always reflect the actual size and number of existing and productive plots. Given the country's recent history, rural communities are sceptical of the continuous campaigns local and regional governments have launched to regularise land tenure. They see these actions as a strategy to widen the tax base or gain control to the means of production (personal communication with farmers, Colombia, June 2013). The irregularity of land ownership resonates loudly in the credit market.

In a different study, Lozano studied access to credit. On this occasion, he found that large producers identified high transaction costs as disincentives for borrowing. Small producers, for their part, were mostly concerned about the lack of information (2009). Misinformation creates further barriers to gain access to credit markets. For instance, Lozano found that farmers all too often self assess whether they are eligible or not to apply for loans. Even before approaching a financial institution, they are convinced no credit will be granted. This has dissipated demand for a service they badly need.

In his research, Lozano found that farm size and crops cultivated influenced the probability of obtaining credit loans (2009). Financial institutions expect higher yields to be obtained from larger plots, younger plantations and high portions of land suitable for renovation. Farms with those characteristics consequently obtain more credit, further supporting the author's findings that coffee farmers who have access to credit report yields up to 50 per cent higher than their counterparts. According to the Organisation for Economic Co-operation and Development (OECD), micro-credit has expanded rapidly in the last few years. It has gone from nearly zero in 2000 to represent 0.8 per cent of Colombia's GDP in 2009 (OECD, 2008, 2010). This illustrates the minor role that the banks have played in the country's rural development efforts. When farmers have indeed resorted to borrowing, they have relied instead on the financial services delivered directly by the Coffee Growers' Federation.

Falling productivity

Between 1995 and 2006, productivity losses in the coffee sector were partially offset by an increase in the extension of cultivated land. In fact, for the last 30 years, output gains have been modest despite considerable higher levels of support for the agricultural sector (FAO, 2013). According to the World Bank (Anderson and Valdés, 2008), nominal agricultural support in Colombia went up from −11 per cent between 1965 and 1979, to 2.5 per cent by 1980 and as high as 16 per cent between 1990 and 2004. Such protection was mostly directed to commercial policies and away from input subsidies and domestic support measures.

By 2007, additional measures were introduced to make Colombia's domestic agricultural sector more competitive in light of the progress the country was making towards reaching new trade agreements. The Safe Agricultural Income Programme (*Agro Ingreso Seguro*, AIS) was put in place for farmers to be better prepared for their integration into the global economy. According to the Colombian Agricultural Institute (ICA), in two years of implementation, the AIS financed over 81,000 agricultural projects, benefiting a total of 73,000 small producers. The programme also helped more than 58,262 families to gain access to irrigation services for the 85,000 ha they owned (Departamento de Planeación Nacional (DPN)/National Planning Department, 2008).

In terms of technical assistance, AIS allocated over COP 50,000 million in financial resources to engage 77,000 families in capacity building activities. The Ministry of Agriculture estimated that, in total and to 2009, more than 316,000 families benefited from the Safe Agricultural Income initiative (ICA, 2009). Mejía Cubillos sustains that, although the AIS was relatively successful in terms of improving the competitiveness of the Colombian agricultural sector, it achieved no further progress in terms of employment and rural development (Mejía Cubillos, 2012; Ministry of Agriculture and Rural Development and National Planning Department, 2011). The scheme was also subjected to intense criticism due to mismanagement and corruption allegations. Criminal investigations were raised against programme beneficiaries and functionaries in the Ministry of Agriculture and Rural Development (Mejía Cubillos, 2012).

Also in 2007, the National Committee of Coffee Growers signed the Coffee Policy Agreement (2008–2011). Endorsed by the then incumbent president Mr Álvaro Uribe, the agreement once stood as 'the most ambitious in the history of the Colombian coffee industry' (Roldán-Pérez et al., 2009: 56). It sought to help the coffee trade successfully overcome the global economic crisis by allocating COP 1.4 trillion worth of economic aid. This was twice the sum provided by the previous Coffee Policy Agreement (2002–2007).

The 2008–2011 agreement included a 'Price Protection Contract', an income-stabilising provision that guaranteed a minimum price of COP 474,000 (USD 199.22) per 125 kg bag. It was expected this price was enough to cover the costs of coffee production. Later on, a new price contract option was introduced to allow growers to secure a price for up to 50 per cent of their total anticipated production (Roldán-Pérez et al., 2009). Contrary to what policy makers believed at the time, these measures have not sufficed to mitigate, let alone address, the negative ripples that result from systemic barriers to raising productivity. In 2013, coffee growers staged a coffee strike locally known as *El Paro Cafetero*. This mass rural mobilisation attests to the urge of taking a comprehensive and structural approach to the coffee crisis. In Colombia, like in many other Latin American countries, rural communities have simultaneously driven national prosperity and bred social upheaval (De Ferranti, 2005).

In 2011, and under the government of president Juan Manuel Santos, the AIS was renamed as the Rural Development with Equity Programme (*Programa Desarrollo Rural con Equidad, DRE*). Borrowing heavily from its predecessor, this new version of the rural support scheme was introduced to make the domestic agricultural sector more competitive. It was repackaged as a response to Colombia's progress towards greater integration with the global economy and the negotiation a series of bilateral free trade agreements (Departamento de Planeación Nacional (DPN)/National Planning Department, 2008). The DRE has given continuity to the government's efforts to improve productivity in the agricultural sector, strengthen the livelihoods of farmers and contribute to reduce inequality in the rural areas. It also seeks to support the internationalisation of the Colombian economy by correcting market distortions in order to boost competitiveness in global markets (Mejía Cubillos, 2012).

The programme was designed and is being currently implemented to assist farmers to face new levels of competition in the agricultural sector. This is particularly true since the beginning of the United States–Colombia Free Trade Agreement. The DRE has an annual budget of COP 500,000 million to help up to 100,000 small and medium producers gain access to loans at preferential interest rates. These resources have been allocated to farmers growing basic foodstuffs, export crops or agricultural produce facing intense competition from new trade partners or countries benefiting from lowered entry barriers (Ministry of Agriculture and Rural Development and National Planning

Department 2011; Ministry of Agriculture and Rural Development, 2013). It also offers non-credit incentives to foster irrigation and drainage works. In terms of building human capital, the DRE has allocated resources to improve the supply and quality of technical assistance services in areas such as management and administration, productivity, entrepreneurship, commercialisation and the environment. Only time will tell how effective these efforts prove to be in developing the capacities required for the agricultural sector to make a substantial shift away from subsistence-level economies.

Additionally, the DRE offers farmers the Incentive for Rural Capitalisation (*Incentivo a la Capitalización Rural*, ICR) to encourage the use of capital assets in agriculture and the undertaking of entrepreneurial activities. In place for more than two decades now, this generation of schemes amount to up to 40 per cent of the total investments incurred and administered by the Financial Fund for the Agricultural Sector (*Fondo para el Financiamiento del Sector Agropecuario*, FINAGRO).

The large scope of these government schemes requires accountability mechanisms to be put in place. Colombia's National University is the body in charge of monitoring and evaluating the performance and execution of the DRE. Assessments are being guided by the principles of resource efficiency, efficacy and effectiveness; social equity, associativity and integration; food security; and articulation with other governmental and institutional programmes (Ministry of Agriculture and Rural Development and National Planning Department, 2011). This scheme does support the cultivation of late- and long-blooming perennial crops such as coffee. Nevertheless, despite these institutional efforts to better the living conditions of farmers, and coffee growers in particular, international commodity prices have not allowed coffee producing communities to pay for better standards of living.

Coffee prices hit US $1/lb in 2012, 2013 and the beginning of 2014. Unsurprisingly, farmers have been unable to cover their costs of production. The way they compensate for the lack of resources is by reducing the use of the most expensive inputs, namely labour and the frequency and intensity of fertilisation. Lower fertilisation levels not only reduce yields but leave coffee trees much more vulnerable to rust, other diseases and plagues. Disinvestments in coffee production also occur as land use changes or new crops are introduced on coffee farms. These elements have enormously hurt productivity, yields, total farm output and consequently income.

Some coffee growers have resorted to producing premium-paying specialty coffee to confront systemic challenges. These initiatives are potentially more profitable but as with any long-term strategy, farmers may have to make significant capital, time and labour investments before they see an incentivising pay-off to their efforts. The prospect of placing their crop in premium paying markets naturally draws producers to participate in sustainability-oriented initiatives. Premiums are, after all, straightforward market signals reflecting consumers' willingness to pay higher prices for better quality, more sustainability produced coffee.

The extent to which the links between price and sustainability practices are beneficial to farmers depends not only on the actual prices paid per pound of sold coffee. The volumes farmers are able to sell and these higher paying markets willing to buy from them make all the difference. Volume ceilings have limited the positive impacts of schemes such as Fairtrade. Farmers are paid premiums above market prices but are only able to sell a small share of their total coffee harvest to this initiative (Kolk, 2012; Valkila, 2009; Bacon, 2005). Considering price premiums alone is not a useful indicator of the economic health of coffee producing households and communities.

From October 2012 to February 2014, and without taking into consideration government support, internal and global prices for mainstream coffee remained at least USD 0.20 beneath production costs. Prices have shown a considerable recovery since the second quarter of 2014, closing costs gaps and possibly allowing farmers to break even or make a profit. At previous prices, even Nespresso AAA farmers, who received price premiums attached to the superior quality of their coffee, were barely managing to cover production costs. From July 2012 and for the next 18 months, farmers grew coffee at a loss. The situation marginally improved once government support and subsidies were factored in. According to anecdotal evidence, Nespresso farmers were using state support to break even (Carlos Rojas, Asoexport, personal communication, 2013). Putting cyclically low world prices aside, farmers' income has been slashed by plunging productivity rates. Output per hectare has suffered a great deal due to plagues and diseases attacking coffee plantations in Colombia in recent years.

For almost a year now, a leaf rust (*Hemileia vastatrix*) epidemic has been creating havoc in all coffee producing countries in Central America. On average, 80.4 per cent of the coffee plantations in Guatemala, Honduras, El Salvador, Costa Rica and Nicaragua grow varieties susceptible to the disease (CATIE, 2013). In fact, rust is so damaging that it is considered to be one of the seven plant diseases leading to the greatest socio–economic losses of the last 100 years (CropLife Latin America, 2013). With an incidence above 50 per cent, affecting around 467,911 ha, this episode of rust is the worst since the fungus was first detected in the region back in 1976 (CATIE, 2013; ICO, n.d.; Rodríguez and Duque, 2009).

The severity of this situation urged affected countries to convene during a special summit in Guatemala, the most badly hit country in the region, during 18–20 April 2013. Attending parties discussed the structural and external factors that have allowed rust to emerge with the intensity and pervasiveness it has had in the last few years. Different groups assessed the situation, possible short- and medium-term control measures and the socio-economic impact of the outbreak on the living conditions of some 505,000 coffee producers and their families. They also discussed mitigation strategies, technical extension models, rust monitoring and early warning systems, as well as communication and training systems (World Coffee Research and PROMECAFE, 2013).

As the situation worsened, a few months later, from 8–10 October 2013, the Tropical Agricultural Research and Higher Education Center (CATIE) in Costa

Rica held the I Regional Forum on Coffee Rust Mesoamerica. During the event, it was concluded that since 2012, and to early 2014, some USD 681.3 million had been lost due to sharp production contractions and forced pruning in 28 per cent of the region's plantations. For the 2013–2014 coffee cycle, it was calculated that forgone income in hard currency for the region would reach USD 161.3 million. More discouragingly still, these figures are likely to be appreciably above the original pecuniary estimates as international coffee prices have gone up. Consequently, the income rust-affected Central American countries could have obtained for this crop season is higher (CATIE, 2013).

Traditionally, the Típica, Bourbon and Caturra coffee Arabica varieties have been planted in Colombia. Known for their excellent agronomic profiles, these trees are vulnerable to *Hemileia vastatrix*, coffee rust. These varietals are characterised by a mild, acidic flavour, a pronounced aroma and moderate bitterness (Forum del Café, 2009). The Típica varietal was first taken to the Americas over three centuries ago via Martinique. Despite being a low yielding plant, it is of exceptional quality and a sweet, balanced, and pleasantly acidic cup profile. Similarly to the Típica varietal, Bourbon is also known for being low yielding. Yet, it is a delicate tree with small, round cherries in burgundy red or yellowish tones. Finally, Caturra trees, the once staple coffee variety in Colombia, grow lower to the ground and are sturdier. Their cherries are red or yellow depending on the strand. When processed, they give a bright and well-balanced type of coffee. These three varietals are highly susceptible to leaf rust and are thus being replaced by the *Variedad Colombia* Castillo®, a higher yielding, rust-resistant plant (Azahar Coffee, 2013).

When the disease first appeared in Brazil in 1970, Colombia's National Coffee Research Centre, Cenicafé, began developing a number of local varieties known as *Variedad Colombias*. Field-tested and handed out to coffee farmers as early as 1982, these rust-resistant hybrids have been improved for more than 20 years to combat blights whilst increasing yields (Rivillas Osorio et al., 2011). Castillo® has proven to be the most successful and most widely planted tree of *Variedad Colombia*. This is due to its resistance to rust as well as to other plagues and diseases. More importantly, this coffee tree type is characterised by higher yields, larger beans (up to 80 per cent qualify as *supreme,* thus giving farmers access to international markets demanding beans of that quality) and good cup quality (Azahar Coffee, 2013; Alvarado-Alvarado et al., 2005). Despite the existence and availability of improved coffee varietals, the lack of financial resources to offset temporarily forgone yields at the time coffee trees are renovated exposes large tracks of land to rust and other diseases (*El Tiempo*, 2010).

Rust first arrived in Colombia in 1983, leading to an outbreak that extended from 1985 to 1987. A second episode took place in 2008, when old plantations and high precipitation levels led to the propagation of leaf rust, causing widespread devastation and considerable plunges in the productivity levels and incomes of thousands of coffee producers (Cristancho, 2012). Two years later, the situation further worsened as phytosanitary problems aggravated. An intense summer resulted in a sharp rise in the levels of berry borer (*Hypothenemus hampei*) in the region of Huila and then the rainy season intensified rust attacks in non-resistant varieties (Rivillas Osorio et al., 2011). Additionally, rust began affecting coffee trees at higher altitudes, younger trees and new plantations (World Coffee Research and PROMECAFE, 2013). This situation led to dramatic output losses. In 2008 production reached 12,515,000 bags. The year after this figure contracted by 30.77 per cent when barely 8,664,000 coffee bags were produced.

Cenicafé, the National Coffee Research Centre, has attributed the latest higher leaf rust incidence to unusually high precipitation levels, diminished sunshine hours from 2006 onwards and aged coffee trees. Old and undernourished plants, insufficiently exposed to sunlight, are 50 per cent more prone to rust than shade-grown coffee trees (World Coffee Research and PROMECAFE, 2013). Certain agricultural practices have been equally damaging. Lower fertilisation levels, as an immediate way to cut down on production costs, uncertain seed origin and lack of regular tree and field weeding have exacerbated the incidence of leaf rust.

As the first response to this problem, the Colombian Coffee Growers' Federation promoted the renewal of coffee trees, this time with rust-resistant varieties (Cristancho, 2012; Café de Colombia, 2011). Starting from 2011 and for the next five years, the Federation envisioned the implementation of an ambitious coffee tree renovation scheme that, upon completion, will have substituted some 130,000 ha every year with improved Colombia and Castillo® varieties. In total, from 2011 to 2015, 150,000 ha of land will have been renovated by pruning, 400,000 ha by cultivation and 100,000 ha of new land will come under coffee plantations.

Prior to that, renovation initiatives had already been intensified in the 1990s but trees were being replaced by more of the same varietal. From 1965–1966 to 1989–1990, 228,000 ha of new land were brought under coffee cultivation but more than twice as many – 521,000 ha – were renovated. This 2.29 ratio of renovated land to new areas of coffee cultivation increased considerably to 10.43 between 1990–1991 and 2009–2010. Between 1991 and 2010, and notwithstanding the more limited availability of land, coffee production did expand to some 85,000 ha. In that same period, 991,000 ha of coffee plantations were renovated in an attempt to control the outbreak of and damages caused by rust. Figures 3.5a and 3.5b depict how the amount of renovated and newly cultivated land changed in the 1965–2010 period. In the long run, it is expected that as the country strengthens its coffee production capacity through renovated plantations with rust-resistant and younger trees, yields and productivity will rise (Café de Colombia, 2011). Colombia's 2013–2014 harvest season has already shown positive signs of recovery.

Cenicafé-led research sustains that rust control and mitigation measures are economically viable operations, especially when compared to forgone harvests and thus income. Research conducted by this centre shows that, on average, rust-affected plantations suffer losses of 23 per cent of accumulated production

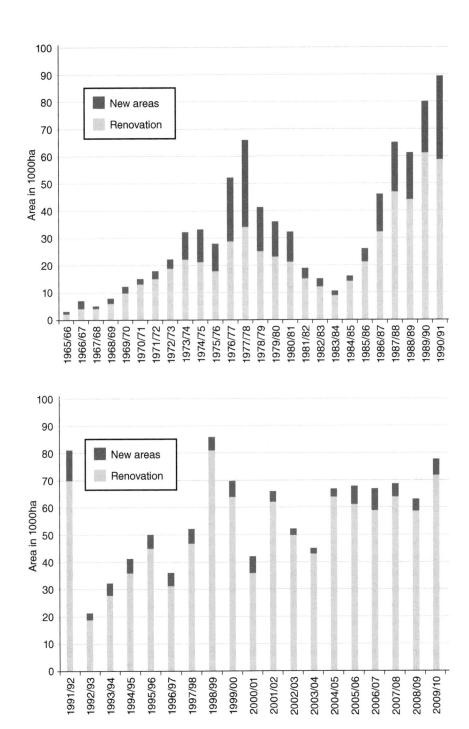

Figure 3.5 New and renovated coffee areas, (a) 1965–1991 and (b) 1991–2010

Source: FNC (2007, 2008, 2011a, 2011b)

for four harvests. Such adverse impacts can be clearly seen in the yields per hectare reported for 2009 and 2010, when rust severely affected Colombian coffee plantations.

In 2009, yields declined to only 8.8 bags per ha compared to 13.2 bags of 60 kg each per ha the previous year. In 2011, the FNC estimated the fumigation costs per ha to control and mitigate leaf rust infections. These calculations were made taking into consideration different types of fungicides and the number of times they are to be applied in a year for trees to be protected and the harvest saved (FNC 2012b). Investments ranged from COP 201,858 to COP 646,157 (USD 100 to USD 320.22 in 2014 prices).[1] Costs varied depending on the fungicide, labour-intensity of its application, the equipment required, as well as other inputs such as fuel, lubricants and personal safety considerations (Rivillas Osorio et al., 2011).

For instance, Cenicafé has directed its efforts to promote a series of recommended practices that will strengthen coffee production in Colombia. These activities include the establishment of production cycles, planting coffee trees of the rust-resistant Castillo® variety, producing plantlets at the farm level, using coffee pulp as fertiliser, setting optimal planting densities according to the productive system, managing weeds integrally, using fertilisers according to soil analyses and applying them throughout the fields. It also recommends integrated management of rust and other plagues, renovating old coffee trees and conserving the initial tree population. Regarding the use of machinery, the institution suggests calibrating sprinkling and coffee processing equipment, selectively harvesting ripe cherries, adopting ecological coffee processing practices and drying coffee beans properly.

On the management and livelihood diversification front, the coffee research centre encourages farmers to produce other foodstuffs and use accounting and cost analysis tools (Ministerio de Cultura, 2011). To supplement coffee income, households grow their own food or other cash crops, such as bananas, plantain, avocado and pineapples. This allows coffee producing families to cover household expenses and carry out recurrent coffee management practices. However, the limited economic and overall benefits accrued from those activities pushes household members to migrate to other rural communities or to urban areas. Even when they stay in their own communities, people seek off-farm employment opportunities. As of late, the boom in the mining industry has been pulling, and keeping, a growing number of workers away from farms.

In 2011 alone, the 'Rust-free Colombia: a National Purpose' campaign (*Colombia sin roya: Un Propósito nacional*) delivered fungicides and fertilisers to 200,000 coffee growers to strengthen and protect some 170,000 ha (Café de Colombia, 2011). In 2009–2010 alone, approximately 72,000 ha of coffee areas were renovated and a further new 6,000 ha were put under cultivation. From 1991–1992 to 2009–2010, more than 990,000 ha of coffee were renovated and a further new 95,000 ha brought under cultivation. In that period of time, on average, for every 1,000 ha of renovated plantations, new coffee areas grew by

95.96 ha (FNC, 2013b). The amount of land under coffee production has thus remained roughly constant. Yields, on the other hand, have declined steadily throughout the decades.

For 13 years, from 1970 to 1983, the area under coffee cultivation surpassed 1,000,000 ha, reaching a maximum in 1970, when coffee was grown in 1,070,000 ha of land across the country. During this period, yields were highly variable, ranging from a low level of 6.8 bags of coffee per ha to a high of 13.5. It took another 20 years for productivity to reach its historical maximum. Yields finally peaked in 1991 and 1992, when output per ha was more than 17 bags. Those productivity levels were nevertheless short-lived and have since steadily declined.

In 2009, for instance, coffee was being cultivated in 920,000 ha of land but yields remained at an average of 8.8 bags of green coffee beans per ha. The following year, the area cultivated increased by 15,000 ha and per ha yields rose to 9.5 bags. In this scenario, the apparently large tracts of new land brought under coffee cultivation are only adding up to reach past production levels (Vallecilla, 2005; ECLAC–FAO, 1959). The (re)introduction of coffee in certain areas has, nonetheless, served to maintain a more or less constant output level rather than resulting in a sharp increase in the number of bags produced. These variations are reflected in Figure 3.6, which portrays both the total area under coffee production and average yields per ha from 1926 to 2010.

World Coffee Research and the Regional Cooperative Programme for the Technological Development and Modernisation of Coffee (PROMECAFE) found that when the price of coffee goes up, the amount of fertiliser used rises (World Coffee Research and PROMECAFE, 2013). As one of the costliest inputs, fertilisers and the extent to which these are used quickly respond to a farm's overall economic status. With fewer resources, coffee producers downsize farm operations and engage in less effective farm management practices. This creates a vicious circle of disinvestment, productivity losses and forgone income. Compared to the value of the industry, investment levels have remained significantly low.

Moreover, it has been observed that farm size is not relevant in controlling crop diseases but that well-managed plantations, with the right coffee tree varieties, are the least affected by rust outbreaks, notably AAA Nespresso farms. Coffee growers working with the premium brand obtain higher harvest incomes. This relatively better economic situation is attributed to participation in the AAA initiative. Farmers gain access to more resources, their plantations are younger and tend to have been renovated with rust-resistant trees and have better access to timely and relevant technical assistance and sector information. These are some of the reasons why AAA farms have been less affected by leaf rust and other plagues and diseases than average plots. Nespresso's sustainability scheme and the impact it has had on participating farmers are discussed in detail in Chapter 6.

Nevertheless, the links between profitability, production, price, climate and technical information urge coffee growers, and the organisations they have created, to take a holistic approach to managing and controlling coffee plagues and diseases. Rust-affected regions obtain lower coffee income, suffer losses in quality and quantity and offer fewer employment opportunities for seasonal workers. The externalities and aggregated consequences take the shape of graver deteriorations of the quality of life of the farmers and more vulnerable rural and coffee economies.

World Coffee Research recommends the use of coffee tree varieties that are both highly valued in the market as well as resistant to diseases and plagues. Thus, prior to renewing plots, farmers and technicians need to be warned about the susceptibility of the varieties generally found in a given cluster or region. As a risk management and diversification measure, farmers can plant a wider range of varieties.

Renewal and pruning are best done as soon as necessary as these activities are clearly important to maintaining long-term productivity levels and good plot health. Undertaking them, however, implies at least partial, albeit temporary, harvest losses producers cannot always afford in the short term. Farmers require support and incentives to improve existing fumigation techniques and pay close attention to size, frequency and timing. Farmers oftentimes know what to do to protect their farms from diseases. What they lack are the resources to do so pre-emptively rather than reactively. They are thus pushed to finding alternative economic activities to compensate for low prices, lost productivity, forgone harvest due to tree renewal, etc. (World Coffee Research and PROMECAFE, 2013).

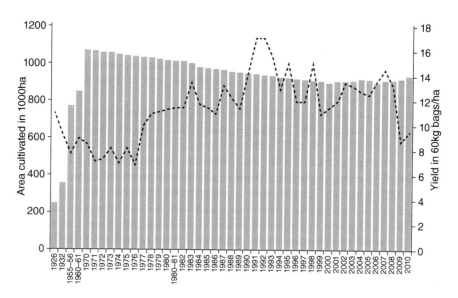

Figure 3.6 Total area under coffee production and average yields per ha, 1926–2010
Source: Vallecilla (2005); ECLAC–FAO (1959)

This generations-worth of knowledge farmers possess is often overlooked. Putting it to use is often limited given information asymmetries about the local and international markets and rapidly changing global supply chains. Production-related decisions are made even more complicated by uncertainty. This is an already complex enough process that variable factors, such as climate change, can only further complicate.

Climate change: making the already variable even more unpredictable

All around the world, with Colombia no exception, climatic variability has always been responsible for fluctuations in coffee yields. Even the slightest alteration in precipitation, hours of sunshine, moisture in the air and soil composition patterns invariably affect the quantity and quality of the agricultural output from which millions of people depend for their livelihoods. Climate change is expected to make the already variable climatic conditions even more so, aggravating existing problems in agricultural production, water management and food security. This is perhaps the reason why global warming and climate change are causing anxiety in the coffee sector (Panhuysen and Pierrot, 2014; Mandell, 2013).

Nothing much is known about how exactly climate change will affect ecosystems, flora, fauna and human activities at local, regional and global levels. Even less is known about the likely extent of the changes and how individual crops will be affected. Expectations, on the other hand, abound. Changes in temperatures are expected to affect planting seasons, render certain producing areas less suitable for coffee cultivation, increase the incidence of opportunistic pests and diseases – such as coffee rust and the berry borer – and intensify the need for irrigation. It is also expected for the hydrological cycle to become more intense. This will naturally impinge on productivity and output as well as put additional pressure on already stressed water resources and water management systems (Panhuysen and Pierrot, 2014).

Climate change is also expected to impact coffee quality. A study by Bertrand and colleagues (2012) found that climatic conditions during bean development affected the chemical composition of coffee seeds. High temperatures, captured in mean air temperature, can induce the accumulation of two alcohols (butan-1,3 diol and butan-2,3 diol) that in turn are closely correlated with a reduction in aromatic quality, acidity and an increase in off flavours, namely earthy and green flavours (Bertrand et al., 2012). Uncertainty will not only affect how much coffee is produced. Quality is probably one of the most aggravating potential results of variations in climate.

The only thing certain is that climate change is leading to irreversible changes in Earth's systems and that impoverished, isolated and small farming communities will be the least equipped to bear a disproportionate share of the consequences. Moreover, since smallholders produce most of the coffee in the world, any additional risk they may face on top of already volatile market pressures will invariably affect where, how much, how and what type of coffee is grown. The 2014 edition of the coffee barometer warns about a substantial decrease in the areas currently suitable for coffee production in Brazil, Honduras, Uganda and Vietnam by as early as 2020 (Panhuysen and Pierrot, 2014). This could cause potential disruptions to trade practices both in this group of very diverse countries as well as at the international level. Compromised output levels and quality of those yields would, as a result, curtail the viability of the sector as a whole, hitting particularly hard those coffee producers, buyers and roasters dependent on highly climate sensitive, highest quality beans.

Like all human activities, agriculture contributes to the emission of greenhouse gases (GHG). This naturally means that coffee production needs to look at mitigation as much as at adaptation measures. This increases the urgency coffee growers face to develop additional skills and capacities as well as to improve farming practices in the short and longer term. It also calls for coffee growers to continuously enlarge and adapt their profile as farmers. In Colombia, the United Nations Development Programme (UNDP) and the FNC are working together to mainstream biodiversity in the coffee sector and support the economic and ecological viability of producing biodiversity-friendly coffee. Operating through public–private partnerships, this 'Green Commodities Programme' has built conservation corridors to connect agricultural plots with forest remnants and planted indigenous tree species (UNDP, 2014). These are encouraging initiatives but the limited scope of current implementation suggests their impact on overall coffee production is likely to be limited or be only fully realised over a long period of time. Scale and quick response is vital.

Nestlé is Colombia's largest coffee buyer. It is not surprising that the company's promotion of structural changes in conventional production practices is generating positive externalities in a large number of coffee producing areas. The existing pool of sustainable agricultural knowledge has gradually expanded, farmers have begun to acquire and put to use entrepreneurial and quality-oriented skills and the quality threshold in coffee production has been substantially raised. Other notable positive spillovers include intensified competition for high quality beans, which exerts upward pressure on coffee prices paid for lots of comparable quality.

Competitors alike stand to benefit from the sustainability initiatives rolled out by this industry leader. In the same way, Nestlé stands also to gain from similar actions undertaken by its competitors. After all, competing industry players are exposed to similar supply risks. The coffee sector as a whole stands to gain from the adoption, implementation, mainstreaming and entrenchment of sustainability-oriented practices. As better agricultural techniques become commonplace and standard practices, the sector's formal and informal institutions and infrastructure will become more resilient and better equipped to face structural and sudden changes alike. Climate change will bring about swift changes. But it is only one of the several issues in coffee production that call for systemic and sector-wide responses.

Succession: where will the new farmers come from?

Currently, coffee farmers are on average 52 to 54 years old. This means growers have around 35 years of knowledge in coffee production but little experience in behaving and acting as coffee entrepreneurs. From our discussions, it was evident that growers are having considerable difficulties in thinking of and managing their farms as successful business enterprises even when they have to sustain families that on average have four children.

Schooling levels are low and attainment deficient. On average, children have completed 4.7 years of formal education. Besides low attendance, the children of coffee producers in coffee growing areas attain lower scores in standardised exams than their counterparts in the regional capital city (Regional Economic Observatory, 2013). In general, rural areas fare poorly in such exams and children there perform well below students in the municipal heads (CRECE, 2013). Given low rates of academic education, it is not so unexpected that less than 1 per cent of the rural population has attended university (Pérez and Pérez, 2002).

Equipped with basic reading, writing and mathematical skills, almost all growers keep mental records of budgets and accounts. They very seldom have a clear idea of their total incomes, expenses and net profits or losses. When income is low, or lower than in previous years, farmers cut down on non-essential goods, services, activities and inputs for coffee production. The flagrant absence of recorded estimates of the economics of coffee farms limits the knowledge and capacity farmers have to formulate long-term plans. They lack the skills to make their farms progressively more profitable, which in turn limits the resources available to them to offer a better quality of life to their families and make coffee farming an attractive profession for their children.

Under the current conditions, and drawn to other booming economic sectors such as mining or the service industries, it is exceptionally rare to hear about rural youths attracted to working the land and cultivating coffee. Consequentially, in spite of the relatively high fertility rate in rural areas and coffee producing regions, generational relay is not assured. The average coffee farmer is relatively old. The young are moving to urban areas where choice of employment opportunities seems broader, public services and amenities are more widely available, recreational and entertainment options abound and life is perceived to be less hard than on the farms.

Toro (2005) describes how grandparents are heading an increasing number of households. They, along with other relatives who are members of extended family structures, care for the children and adolescents of parents who have migrated to urban centres or abroad in the quest of better-remunerated or more socially recognised employment opportunities. He argued that due to the coffee crisis, family, as a social and economic institution, has normalised and accepted migration as a resource that can bring support and benefits for all. The combination of this exodus with ageing households and farmers slows down the rate at which agricultural, technical and technological innovations are adopted (Perdomo and Mendieta, 2007; Lozano, 2009; Nates Cruz and Velásquez López, 2011). Technically and technologically outdated and environmentally and economically unsustainable local practices fare poorly in a dynamic global economy. They tighten the unprofitable cycle in which many coffee producers are caught.

The previous section on ageing factors of production made reference to the linkages Lozano (2007) found between household conformation and production levels. The age of the head of the family has negative effects on production. Education, even at the most elementary levels, on the other hand, has a positive impact. Perdomo and Mendieta (2007) made similar findings. Their study confirms claims that the younger the producer and the more education he has received, the higher the productivity will be. This is explained by the greater adaptive ability of young farmers. They are quicker and more effective in adapting to and adopting technological innovations in agriculture. In Colombia, an aged and ageing coffee workforce implies that essential changes to production will take longer to arrive, assuming they ever do in the first place.

As early as 1993, emigration-induced labour shortages were recorded in the coffee-growing regions of Caldas, Risaralda and Quindío. Similar observations were made in Huila, Nariño and Cauca, although to a lesser extent. That year, out of the 561 coffee-producing municipalities, 109 experienced labour shortages of various degrees. These areas represented 19 per cent of the country's total land under coffee cultivation and 52.8 per cent of that year's total coffee production. Acute labour shortfalls were felt in 24 other municipalities, moderate deficits in 44 more and low shortages in the remaining 41. By 2005, the number of municipalities with a deficit of available labourers increased to 22 per cent, from 109 to 124. These places accounted for 55.5 per cent of total coffee production. Compared to 1993, four more municipalities faced acute shortages putting the total at 28. Regions confronted with moderate and low labour insufficiencies also rose to 55 and 41 municipalities respectively (Leibovich and Botello, 2008).

The lack of workers was explained primarily by rural to urban migration. Shortages were also due to labour movements from coffee growing areas to other rural regions where more profitable cash crops are being produced or other economic activities are proliferating. Labour insufficiencies indicate the urgency to consider ways to rapidly respond to demographic changes, the shrinkage of the labour pool in coffee producing areas and the need to increase productivity for whatever labour is available.

Between 1993 and 2005, Caldas, one of the areas from where Nestlé currently procures green coffee, witnessed one of the lowest national population growth rates at 0.24 per cent. The region also recorded the highest emigration rate in all of Colombia: 8.59 per 1,000 inhabitants. In Antioquia, Caldas, Quindío, Risaralda and Tolima, the cohort comprising people between 25 and 35 years of age diminished significantly, a decrease mostly attributed to rural to urban as well as to international migration.

According to the 2005 National Census, between 2001 and 2005, 67.9 per cent of international immigrants leaving Colombia previously lived in coffee-growing municipalities. In the coffee regions located in central Colombia, the productive share of the population fell more than proportionally to the non-productive population cohorts (Leibovich and Botello, 2008). These demographic shifts are partly explained by the country's new economic profile and perception of improving political stability and partly by development patterns. Urbanisation, after all, has historically accompanied social and economic development.

In a historical narrative of rural outmigration, Nates Cruz and Velásquez López (2011) stress that these outflows of people are not uniquely related to current low coffee prices or the rise of other productive activities in rural areas, such as mining. In fact, population movements away from coffee areas and to other rural communities and cities have more to do with the value attached to agricultural activities. During the coffee bonanza, the children of well-to-do farmers could afford to move to the cities or abroad to attend school and university. After all, younger adults have always been more likely to migrate. In the vast majority of those occasions, sojourners became migrants. They did not make it back to the coffee farms. In the aftermath of the coffee crisis, movements to the cities only intensified. Little to nothing could be done to tie youth employment to unpromising economic activities.

Traditionally, people have been constantly presented with a representation and narrative of progress and prosperity that mostly excludes agriculture and residence in rural areas. This dialectic is not exclusive to Colombia. It is recurrently found in developed and developing countries alike. To the vast majority of rural dwellers and people in agriculture, the fields offer very limited possibilities of ascending the social and economic ladder. In response, they have historically moved and sought better opportunities in urban areas or wherever the distribution of resources, services and opportunities is less unequal. They look for places where mobility can be attained in a not so distant future.

In addition to these socio-economic push factors, in Colombia, people have also moved to escape violence. Large numbers of migrants and around 86 per cent of the conflict affected internally displaced persons (IDPs) have moved to towns and cities where they encounter limited employment possibilities given their low levels of academic attainment. The Consultancy for Human Rights and Displacement (CODHES) estimated that, between 1996 and 2001, almost 2.9 million people were internally displaced. Some of the most affected areas are coffee growing regions, comprising of municipalities where Nestlé factories are located, the Nescafé Plan is under implementation and from where Nespresso sources some of the brand's beans (Pérez and Pérez, 2002).

The country is undergoing significant changes in its economic structure. Despite the focus the government of President Santos has placed on economic growth and reductions in inequality, the oil and mineral resources boom and large capital inflows have led to an appreciation in the exchange rate of the

Colombian Peso. Higher GDP values have had two opposite effects. The discovery of capital intensive and natural resources in the extractive industries has eroded price competitiveness, hurting exporters and non–resource industries. Concurrently, as the currency has strengthened, imports have intensified as they have become relatively cheaper (Ebrahimzadeh, 2012).

This economic phenomenon, known as the Dutch Disease, is undermining the competitiveness of the Colombian agricultural sector. Currency appreciation has further eroded profitability in the primary sector, which particularly affects coffee growers as well as farmers cultivating cash crops. As a response to an increasingly uncompetitive and impoverished rural sector, and other structural factors such as low labour productivity, fewer young people are joining the next generation of coffee farmers. The study team did not find a single farmer during the field visit whose children wanted to continue with the family tradition. In fact, young adults were nowhere to be seen. Colombia's coffee hills were being tilled and harvested exclusively by older men and women.

Young people have been traditionally attracted to cities. They are drawn to the economic promises they apparently hold, the amenities and entertainment options that seem to be more widely available, the possibilities of social advancement and the higher levels of excitement they expect to find there. Urban regions are appealing to the young but this appeal is only a part of the urbanisation story. In Colombia, as is also the case in many Latin American countries, urban growth and migration to cities seems to be more the result of a crisis in the rural sector than urban-based development. Rural youth often aspire to a different future and seek employment outside agriculture, the coffee sector not being an exception. The young are pulled into cities or larger settlements by images of success and the prospect of closing perceived gaps in income, access to services, entertainment and quality of life separating life in urban and rural areas.

Migration to cities is thus fuelled by development and wealth gaps dividing rural and urban areas. These differences are likely to continue, unless specific and effective policy measures to generate prosperity in the countryside are formulated and implemented. In the 2005 National Census, 4.4 per cent of the working age population stated having changed their place of residence at least once in the last five years looking for jobs. They had also moved due to the lack of economic opportunities in their home communities. According to those figures, 4.9 per cent of Colombian men have moved to a different place in the country in the pursuit of better labour options. This figure was 3.9 per cent for women (Mejía, 2011).

Despite the high poverty rates in urban areas, estimated to be around 40 to 50 per cent, income levels are still higher compared to rural areas. There, poverty rates range between 60 and 70 per cent (FAO, 2013). Colombia's economic, social and environmental realities suggest that migrants, many of them young people from farming families, may not find in the towns and cities the opportunities and the higher quality of life they are looking for. Most of urban unemployment has a rural origin.

According to Botello (2010), urban wages paid to unskilled workers are determined by wages paid in eight coffee-producing areas in the country: Antioquia, Caldas, Cundinamarca, Huila, Nariño, North of Santander, Santander and Valle. Her findings suggest that the urban and rural labour markets are interlinked, particularly to the coffee sector. This finding may have important consequences for those coffee labourers moving away from agricultural activities in search of better salaries and quality of life in bigger towns and cities. Since rural and urban prosperity are interrelated, the rural sector ought to be made competitive and worker productivity and wages higher in order that salaries in non-agricultural activities can also improve. Economic models looking at dual economies further support these observations.

Economist Arthur Lewis-inspired theories of dual economies point at the allocation inefficiencies that lead to stark productivity gaps between two broadly divided sectors: rural (traditional) and urban (modern) industries (Lewis, 1954). According to McMillan and Rodrik (2011), labour reallocation from less to more productive activities leads to economic growth without necessarily increasing productivity within sectors. In Colombia, labour has moved to less productive, and mostly informal, activities. It has been estimated that 50 to 70 per cent of the labour force is employed in the informal sector. Out of the unregulated workers, 40 per cent earn 1.2 times the minimum wage. Unemployment also contains a sex and age bias, mostly affecting females, young (aged 15 to 24) and older people (OECD, 2010).

Moreover, the current boom in large-scale extractive industries is generating growth but not enough vacancies to absorb all the people who are joining the labour market or moving away from agricultural activities. In 2010, agriculture stagnated whilst mining grew by 11.1 per cent (FAO, 2013). The next year, mining alone accounted for 24.2 per cent of exports, contributed 2.4 per cent to the GDP and received 20 per cent of the foreign direct investment (FDI). In 2011, over COP 650,000 million were spent on the construction of infra-structure aimed at fostering this extractive industry (SMGE, 2013). Whilst it is true that labour productivity is much higher in mining and natural resources than in agriculture, these activities absorb a limited number of workers. For example, mining and oil employ only 1 per cent of the labour force. The shrinking industrial and agricultural sectors, in comparison, create 13 per cent and 17 per cent of national jobs respectively (DANE, 2013).

Employment creation is an important way to reduce poverty. But the extent to which poverty will be reduced depends on the type of jobs created. Ideally, jobs should be formal, stable, socially recognised and profitable. This is a labour market transformation to which the private sector can be of great assistance. As part of a prosperous community, economically better-off families can lead healthier, safer and better lives. Additionally, if access to improved social services and modern amenities can be secured, this will certainly encourage some younger farmers and young people to stay in rural areas.

As an example relevant to the case presented in this book, Nestlé was contributing to poverty alleviation in Colombia in at least two ways relevant to coffee production. Firstly, to September 2013, the Swiss company was directly employing more than 2,360 people and a further 1,738 workers indirectly. Secondly, through its sustainability-oriented initiatives, Nestlé is contributing to making coffee production a more attractive, profitable, dignified, competitive and professional economic activity. These activities are described at length in the following chapters.

Yet, it would be disingenuous to assert that these actions alone will be effective at a large scale or will put an end to this high rate of rural to urban migration. In fact, any measure directed at slowing down emigration that does not make a concerted effort to include the input of the youth will have limited success at best. So far, the youth have not been included in the decision and policy-making processes that aim at retaining as many of them in rural communities as possible. Younger coffee growing generations ought to be proactively encouraged to identify the mechanisms, besides just economics, that can render the rural sector, and coffee production, more attractive.

A large exodus of coffee growers away from their plots and into the cities or other economic activities will invariably affect rural labour markets, knowledge about coffee production and output levels. Global and local coffee supply chains also stand to be impacted. Already in 2008, the Sustainable Markets Intelligence Centre (CIMS) pointed at the challenges Nespresso and other Arabica coffee purchasers could expect from these large-scale occupational and physical movements away from coffee production and agricultural activities.

In more affluent countries, the movement of people to cities has historically transformed societies, and fuelled development and industrialisation. This very same process of urbanisation is simultaneously disruptive and transformative of rural and urban dynamics. These forces at play open up new opportunities to revitalise the coffee sector and agricultural activities at large. Moreover, assuming the country will move towards building lasting peace and economic growth will continue to pick up momentum in the future, historical evidence indicates that rural to urban migration is likely to intensify even further.

Already, the children of many coffee growers have moved to large towns seeking employment, especially in the expanding extractive and service industries. This trend is already visible and quite widespread. Mobility, higher expectations and new options have enlarged the array of occupational and residential and lifestyle opportunities for people in the cities and the countryside alike. It is unlikely that these trends will reverse in the foreseeable future. Lamenting the movement of people from one place to another will be of little use. Instead, communities and the public and private sector alike need to think of formulating the right policy mix, developing innovative solutions and guaranteeing the prevalence of enabling conditions for Colombian coffee farmers to turn challenges into potentially beneficial prospects.

For instance, rural outmigration and rural to rural movements can open up opportunities for land consolidation and the creation of larger farms that can better capture the benefits of economies of scale. A modernising coffee sector will probably require fewer people working in it (Nestlé SA, 2004; ITC, 2011;

ICO, n.d.). Anticipating the likely areas where these trends are to intensify, Nestlé, and other private and public stakeholders, could consider encouraging financial institutions to assist farmers in securing the means of production they require. For instance, they could extend credit lines to those coffee growers seeking to acquire neighbouring plots, thereby increasing land size to more optimal levels.

Transportation infrastructure presents another huge opportunity. Currently, the lack of roads is limiting people's mobility and raising freight costs. Colombia's road network is still less than 200,000 km long, out of which only 15 per cent of all roads are paved (OECD, 2010). Consequently, road freight costs are so high that for many, delivering products to the marketplace becomes economically unprofitable once transportation costs are included. For example, during the study group's visit, a group of farmers from the municipality of Sevilla mentioned the idea they came up with to secure cheaper foreign fertiliser through the local co-operative and bring it to Colombia by boat. This should have allowed coffee growers to purchase a key agricultural input at more reasonable prices. However, internal transportation costs from the arrival port to the points of distribution proved to be simply too high to make it an attractive economic proposition. Delivering basic infrastructure can take the rural sector really far.

This situation is improving, albeit slowly. As more paved roads, railroads, ports and airports are constructed and they become more reliable and less weather-dependent, the transportation systems linking urban areas to the scenic coffee-producing hills have also continued to improve and intensify. This has resulted in the appreciation of land values in the country's picturesque coffee-growing belt, locally known as the *Eje Cafetero*.

In the past, especially during the 1980s and 1990s, drug lords purchased large and prime tracts of land in some 409 municipalities, around 42 per cent of the total, at prices well above market rates. This led to a concentration of productive land, which has been put to little or no use. The forced displacement of thousand of farmers that settled in impoverished peri-urban areas contributed to disinvestment in the sector. In recent times, higher land value is no longer related primarily to what these plots can produce. Instead, tourism has emerged in the recently United Nations Educational, Scientific and Cultural Organization (UNESCO)-protected coffee cultural landscape. Additionally, alternative manufacturing activities have slowly begun and increasingly affluent urban dwellers have intensified the demand for a second home in the countryside (IGAC, 2013; Procaña, 2013; Ministerio de Cultura, 2011).

This cultural and productive landscape was recognised for the tangible and intangible elements that have resulted in a unique geographical, social and cultural legacy (World Heritage Centre, 2008; CRECE, 2007). Family-level, inter-generational and historic employment has been singled out as some of the exceptional values defining Colombia's Coffee Landscape (CCL). For more than a century, families have worked in small farms established along the steep hills of the Andes Mountains. This communion of household labour, challenging geographical landscape and small-size landholdings is one of the area's defining traits and key element to promote the CCL as a World Heritage Site by UNESCO.

The inclusion of the Coffee Cultural Landscape of Colombia (CCLC) in the UNESCO World Heritage List in 2011 has boosted the area's tourist and recreation profile. Every year, it has been estimated that between 450,000 and 500,000 national and international tourists visit a coffee-themed park in Quindío (National Coffee Park, n.d.). In 2010, the so-called Coffee Triangle – an area bounded by the regional capital cities of Armenia (in the department of Quindío), Manizales (in Caldas) and Pereira (in Risaralda) – contributed to 3.8 per cent of the national GDP. In these three departments, tourism, construction and hospitality have begun to emerge as important economic activities. Other sectors with considerable growth potential in the area include the development of the car industry, cosmetics, cleaning products, hotels, tourism and outsourcing services. Between 2007 and 2012, the Coffee Triangle received USD 113.1 million in accumulated investments (Proexport Colombia, 2012).

Efforts to develop agro and eco-tourism as a new economic sector in coffee producing areas have emerged as a promising activity for young people, especially visits to coffee farms. Nevertheless, changes in land use can negatively affect coffee production. If land use changes significantly from agricultural to industrial activities, and land values soar, farmers may be pushed out of their properties, to the detriment of coffee production. In such a scenario, young and future coffee growers alike may not be able to secure additional plots and thus land consolidation opportunities may not be seized. Accordingly, the potential to enhance productivity in the sector through economies of scale may not be realised.

Women

This book uses the gender-neutral terms of coffee growers, farmers and producers to refer to the men and women that work on the many small plots where Colombia's coffee is grown. Regretfully, women's participation in the sector has been largely neglected even when they constitute around 27 per cent of coffee growers. Well beyond their role as members of the rural labour force, women are coffee growers themselves. They legally own 24 per cent of land titles and produce 24 per cent of all technified coffee (Coffee Growers Committee of Caldas, 2012). These numbers suggests that women's participation in the sector seems to be formally recorded. The current FNC president, Luis Genaro Muñoz Ortega, claims that the disaggregated statistical data pooled in the organisation's information system (SICA) can be used to draw attention to women's activities in the sector. However, except for the work of a few counted scholars (Chacón, 1990 and Garzón 2002 in Ramírez-Bacca, 2011), studies describing and assessing women's involvement, and that of family cores, in coffee production are largely absent (Ramírez-Bacca, 2005, 2011).

More generally, mainstream research does not explore the role family relations and women play in coffee production. This lack of information is especially regretful given that the family is the unit of rural economies and its members work towards the economic prosperity and wellbeing of the household as a whole (Aristizábal and Duque 2008a, 2008b). Ramírez-Bacca (2011) argues that rural domestic dynamics is a largely unexplored area for agriculture in Colombia as a whole, to which coffee is not an exemption. Although there are close to 500 municipalities and 19 regions where coffee is cultivated in the country, little documentation exists about the general structure and overall relations found in some 500,000 coffee growing families.

According to Gabriela Silva, General Manager at the FNC, one-fifth of all farms were administered by women. In 20 per cent of those cases, women were also the main breadwinners (FNC, 2012d). The tasks they undertake out in the fields, whether they own them or not, are no different to those carried out by their male counterparts. The coffee sector is not excluded from the double burden women face in virtually all the productive activities they perform. Quite the opposite, they are expected to work the land and at home, taking care of their households and families. Unsurprisingly, women's participation all throughout coffee's productive chain is neither fully recognised nor remunerated. Institutionally, just recently, the Federation has started to address the low levels of female participation in formal decision-making mechanisms.

In early 2008, the FNC organised a series of workshops in different coffee growing municipalities for women to carry out a needs assessment and identify ways to address those issues (Coffee Growers Committee of Caldas, 2012). These participatory meetings revealed important gaps where local and regional governments, coffee institutions, knowledge-disseminating organisations, financial institutions and social associations can step in. There is no need to create new groups, associations or institutions for women to raise their concerns. These avenues already exist, but should be strengthened, recognised, funded and promoted. Instead, participants identified the importance and practicality of mainstreaming gender as a crosscutting issue in already existing schemes and among already engaged partners.

Women also listed the limited coverage and deficient quality of public and social goods as factors limiting their own development in the coffee sector. Regarding physical infrastructure and public service delivery, poor housing conditions, deficient sanitation and health services and the state of disrepair of most roads were listed as important structural challenges limiting growth in agriculture and the development of additional productive activities in rural areas. Other issues raised included the growing demand for social services for the young and the elderly. Regional governments were thus asked to increase support for women's associations and generate inter-institutional spaces where gender specific issues are addressed and efforts to contribute to women's empowerment are channelled.

Socially, some of the challenges identified during the workshops were not substantially different from those known to be affecting the sector as a whole.

Lack of employment opportunities for the youth, emigration out of coffee growing regions and school absenteeism are recurrently mentioned as the most insidious problems requiring solutions. Other pervasive challenges included difficulties to access credit and other financial services (Coffee Growers Committee of Caldas, 2012).

Not surprisingly, given the salient organisational and institutional profile the Federation retains in Colombia's coffee growing communities, workshop participants wished to see a more gender responsive and inclusive organisation. Previous efforts seem to have been timid. For instance, in 2005 the FNC created the Programme for Women Coffee Growers as an institutionalised effort to mainstream gender in all of the organisation's activities. With this initiative, the FNC said it is looking to develop the skills women currently lack to achieve greater political representation.

During the 2008 workshops, the FNC was asked to outline the role of women in the programmes it offers. Participants also called for the delivery of training programmes targeting them and their families. As an initial response, between April and November 2008, some 78 Participatory Councils of Coffee Growing Women (CPMC) were created. Such local bodies are believed to open up spaces at the municipal level to promote the role and participation of women in the sector. Some 100,000 women have a FNC ID card. This represents around 24 per cent of all unionised members (FNC 2012d). To September 2014, 552 women were participating on Municipal Coffee Committees and 13 more were active on Regional Committees (FNC, 2012d). Expectations are set for women to be able to influence and make the decisions that impact their lives as coffee growers. Although crucial to achieve gender equality in the sector, fostering and strengthening women's political participation in coffee institutions, as voters and candidates, is only the first step.

A 2012 document made available on the website of the Committee of Coffee Growers of Caldas mentioned a series of actions the FNC had undertaken to advance women's rights. The listed efforts included financial support for the Family with Rural Well-Being Programme amounting to COP 3,885 million, strengthening associations of women coffee growers, sponsoring the participation of ten women leaders in the coffee sector to help build a National Public Policy on Gender Equality, and assisting the Observatory for Women and Coffee Growing Families.

Regretfully, this or similar initiatives were never mentioned during the field visit carried out in May 2013 and thus no primary data on their impacts was recorded. The names, actions and results of such projects were totally absent from group and individual conversations held with women in their different capacities as farm owners, co-operative employees, technical advisors and cluster representatives. In fact, the study team came to know about such gender-directed efforts only after exhaustive literature reviews and deliberate efforts to address that information gap in the research project.

The reality we observed in the field and what literature, and the lack of it, points to the more and greater obstacles women face to make their lands productive and adopt more sustainable practices. These shortcomings persist despite the contributions women make to coffee production. Even when, unlike their counterparts growing other crops, women coffee growers may not be 'statistically invisible', they still receive low or no pay for their work, they have fewer education opportunities and have more limited access to resources such as credit and technical assistance.

The coffee sector is being affected by micro and macro structural challenges alike. Women, on top of it, face social, economic, cultural and political obstacles that further limit their capacity to prosper as coffee growers. Naturally, efforts towards achieving gender equality in coffee production cannot be insulated from broader and more deeply engrained social attitudes. Once again, this stresses the responsibility the government ultimately has to work, alone or with partners, to build a physical, governance, security and institutional framework for farmers, women and men, to lead better lives.

Colombia's coffee architecture

Colombia's coffee institutionalism is unique. Over the course of almost one hundred years, the sector has been represented, organised, structured, supported, assisted, educated, trained, guided, funded and encouraged by a strong institutional and social architecture. The Colombian Coffee Growers' Federation (FNC, *Federación Nacional de Cafeteros de Colombia*) was founded in June 1927 during the II National Congress of Coffee Growers in Medellin. It was the result of the ever more prominent role coffee was playing in the country's development and economy (Almacafé, 2010). It was also a collective reaction to an impending economic crisis in the sector. International coffee prices had weakened and many major traders had either gone bankrupt or left Colombia (Thorp, 2000).

Immediately after its creation, the FNC became a key institution for the development of the sector in the country. Since 1927, it has organised the internal coffee supply chain and represented the interests of producers at the national and international level. For close to nine decades, the exercise of 'coffee diplomacy' has determined many of the incentives and constraints faced in coffee production.

In general, the FNC protects and promotes Colombian coffee domestically and abroad; oversees research activities and coordinates the farm visits of over 1,600 technical assistants. It also ensures low transaction costs for coffee growers through the purchase guarantee and offers information and technology training as well as other education and capacity building programmes. It implements value added strategies for certified coffee; promotes local brand Juan Valdez® and Emerald Mountain®; and makes social investments by leveraging resources and establishing public and private partnerships (FNC, 2012a, 2012b). Amongst coffee producer countries, Colombia is the only one with an institution organising the sector domestically and representing it internationally.

Through these actions, the Federation has built and engrained a sense of ownership amongst coffee farmers. It has historically advanced the development and defence of the sector and filled in state vacuums, in timing and presence. For a long time, the lack of adequate public institutions at the local level created governance gaps that allowed Municipal Coffee Committees to play wider roles, wherever they existed. That way, at the field level, the FNC has made social investments in coffee growing areas, built professional capacities and skills relevant to the sector, and positioned Colombia in the international coffee market. At times when rural violence escalated, its institutional presence partially shielded coffee growing regions from the tensions and fighting experienced in other areas of the country.

One of the most relevant services the FNC provides to coffee growers is the purchase guarantee. This mechanism is a warranty that the Federation will purchase the totality of their harvest at a fair and transparent price, based on current international prices. Prior to its creation, many small farmers received a mere fraction of the price paid for their crop. Intermediaries and speculators used to take a large cut. Once the FNC was founded, it began purchasing coffee from small producers, which raised the domestic coffee price. Farmers then started receiving gradually higher percentages of the actual rates paid for their coffee (Thorp, 2000). This mechanism currently transfers between 90 and 93 per cent of international prices of Colombian coffee directly to the producers (FNC, 2014b) and constitutes only one of the many elements that are part of the safety net protecting them.

The National Coffee Fund (*Fondo Nacional del Café,* FC) was created in 1940 to harness the resources and profits from coffee exports. These resources were to finance the implementation of programmes of shared interest for those in the coffee trade. As initially conceived, the Fund sought to harmonise domestic production with internationally set export quotas by absorbing excess coffee production. Between 1962 and 1989, as Colombia took part of the quota pact reached between exporting and importing countries, the Fund came to operate as a regulating body. It controlled coffee income and supply at times of high international prices. With time, the FC became involved in the delivery of public services in coffee growing regions, funding research and development activities, providing credit and investing in alternative livelihoods (Cárdenas and Junguito, 2009).

Borrowing from the way collective saving schemes are structured, the FC obtains its resources from a tax levied on all exports. USD 0.06/lb are obtained from sales of green coffee, USD 0.48/lb for soluble coffee, USD 0.36/lb for extract and USD 1.08/lb for roasted coffee. The collected resources are used to stabilise and promote coffee production in Colombia and finance FNC's technical assistance efforts, scientific research and quality-control programmes. A part of the funds is invested in social schemes that can improve the living conditions of the country's coffee farmers. Additionally, the FC allocates resources to market Colombian coffee abroad. In 2011 alone, the fund invested an estimated USD 365 million in these activities (Andrade et al., 2013).

The National Coffee Fund's activities are supervised and monitored by the Coffee Growers' National Committee. This is a body formed by high-level government officials from the Ministries of Finance, Agriculture, Commerce and Industry and the Planning Department as well as by representatives from coffee producers' organisations (FAO, 2013). Once fully independent, the FC lost its stabilising capacity after the end of the 1989 International Coffee Pact. Short of resources, it was forced to seek financial support from the national government in 2002. At present, the national coffee policy is just another component of the country's tax and revenue regime (Cárdenas and Junguito, 2009).

After the FNC and FC were established, additional bodies were created to support farmers in marketing their coffee. On 12 November 1929, Almacafé was founded as an enterprise to unconditionally buy, store, and handle exports of Colombian coffee (Almacafé, 2010). In 1985, the Colombian Coffee Growers Cooperatives Export Corporation, Expocafé, began operations. Ten years later, the FNC and the Coffee Growers' Departmental Committee of Quindío inaugurated the National Coffee Park. This public space is dedicated to preserving cultural and historical heritage of the coffee growing areas of Colombia and promoting tourism in the region (National Coffee Park, n.d.).

In 2002, the FNC created Procafecol, a company operating Juan Valdez® retail outlets to gain access to the high-quality, higher-value premium segment. Indirectly, this entrepreneurial endeavour brings about additional revenues for the National Coffee Fund and FNC members, who are coffee growers themselves. The Federation controls the entire production chain for the coffee sold in those establishments and collects royalties from the use of the brand (Andrade et al., 2013). Initially, Juan Valdez® shops existed exclusively in Colombia. After a few years, branches have been opened in 12 countries: Aruba, Bolivia, Chile, Ecuador, El Salvador, Kuwait, Malaysia, Mexico, Panama, Peru, South Korea and the United States (Juan Valdez, 2014). It owns 83 per cent of Procafecol, small independent coffee growers own 4 per cent and the International Finance Corporation (IFC) owns the other 17 per cent. Some 18,000 coffee growers currently own shares in Procafecol.

As noted earlier, by the 1930s, coffee was by far the most important agricultural crop for Colombia. In 1938, the FNC established the National Coffee Research Centre, Cenicafé, in Chinchiná, in the key coffee growing region of Caldas. The institution was set up to study issues related to on-farm production, harvesting, coffee processing, bean quality, agricultural by-product management and utilisation, and environmental. Cenicafé has focused on developing competitive and sustainable economic, environmental and social technologies for the production of coffee in Colombia. At present, scattered over the three mountainous regions where coffee is produced in the country, Cenicafé also has eight experimental research stations gathering representative farm data (Cenicafé, 2014).

The formal and informal institutions coffee growers have created throughout the years have made 'systematic efforts' to facilitate the development of human capacities, research and co-operation that can bring benefits to the sector (Ministerio de Cultura 2011). The institutional arrangement described in this section has thus been crucial for the development of the coffee sector in Colombia. It is now seen as a fundamental, almost indispensable, part of the country's coffee culture and landscape. The Federation has historically played an active role in organising the internal coffee supply chain. It has also exercised a great deal of 'coffee diplomacy' by representing the interests of producers at the national and international levels as well as by influencing public policies. It has financed public projects in coffee areas and invested in infrastructure, education and the delivery of public basic goods and services. Beyond this, the FNC is best credited with having organised the crop's internal market via an extensive network of co-operatives and municipal and departmental committees.

Coffee's architecture and rural support network is thus robust and firmly grounded. Disseminating best agricultural practices as a strategy to attract coffee sellers does not necessarily constitute a competitive edge. In other countries, especially in underdeveloped rural communities, technical assistance constitutes one of Nestlé's most valuable and effective services. More than that, it is one of the company's sources of competitive advantage. When making procurement decisions, farmers have often factored in technical assistance, and not only price, when determining which company to sell their goods to (Biswas et al., 2014).

The reality is quite different in Colombia. There, coffee growers are part of a trade where, for many decades, the formal institutional infrastructure has looked after farmers' interests and the Federation has pursued a wide range of social goals. This has become the *de facto* way in which the coffee business is conducted in the country, thereby shaping farmers' expectations attached to the entry and engagement of new partners. Many farmers now believe, and are aware, that if a company like Nestlé or any other is going to source coffee from them, it should be willing to help them make their social, economic and environmental practices more sustainable. These are improvements not limited to coffee production. They bring benefits that extend to the community.

Farmers working with Nestlé have voiced and raised their expectations regarding positive technological, managerial, technical and skill spillovers (personal communication with farmers in Sevilla, Buga, Valle, Manizales, Supía, Riosucio, Jardín and Rionegro, June 2013). With time, notions of company–society interdependence, which Nestlé calls 'Creating Shared Value', have become more widely and more intensively internalised. Most producers do not necessarily recognise such practices under that, or any other term for that matter. They are, however, beginning to talk about their participation in supply chains and their desire to take a more proactive role. Other changes are also noticeable. Large companies are no longer being referred to as ghost-like figures or remote entities. Farmers have realised that the coffee beans they produce are in fact inputs for coffee drinks consumed all around the world. Global dynamics are being tied to local narratives and local ways of organising production systems.

Peace wears yellow

Colombia's long-established, well-respected and extensive institutional coffee infrastructure delivers a large number of services to the 563,000 coffee producing families in the country. A significant portion of its reputation rests on the agricultural extension services and technical advice the Federation has provided for decades. Yellow-t-shirted agronomists trailing the Colombian coffee landscape have become a sign of neutrality, solidarity, respect, knowledge, support and help.

Social infrastructure is endogenous, determined by history, location, language, culture and politics. Its most efficient social infrastructure provider is supposed to be the State (Hall and Jones, 1999). However, in places where governmental physical and institutional presence has eroded, other actors, legitimate or not, have usurped its place. Fortunately, in many coffee growing regions in Colombia, residing public governance was mitigated by the public services and social infrastructure built and shaped by the Colombian Coffee Growers Federation. During the most intense years of the civil conflict, municipal coffee committees compensated for the lack of effective rural institutions and brought a sense of stability to coffee producing areas.

To this day, the presence and public goods the FNC still finances and delivers affect economic decisions, capital accumulation, productivity, educational attainment and social cohesion. For instance, the Federation has leveraged its resources and position to obtain the support of public, private, civil and international actors to execute different development projects in coffee growing areas. Since 2002, these actors have jointly built some 2,454 km of roads, 4,996 aqueducts and 3,137 classrooms. From 2002 to 2010, 84,809 km of tertiary roads were upgraded. Additionally, since 2003, 31,000 houses have been improved and 328 hospitals and community health centres built. To date, 109,000 coffee farmers have been given access to health care (FNC, 2010, 2011a, 2012d, 2013b).

Nationally, the Federation collaborates with Acción Social (an entity channelling national and international resources to execute all the social programmes), the Ministry of Agriculture and Rural Development, the Ministry of Education, the National Learning Services (SENA), Colombia's Solidarity and Guarantee Fund (FOSYGA), the National Roads Institute (INVIAS), the Colombian Institute for Family Wellbeing as well as local governments in the coffee growing regions. Amongst some of the FNC's international partners are the Inter-American Development Bank, the United Nations Development Programme (UNDP), the German Development Bank KfW, the US Agency for International Development (USAID), the Spanish Agency for International Cooperation and Development (AECID), the then German Technical Cooperation (GTZ) now German Society for International Cooperation (GIZ) and Nestlé (both Nescafé and Nespresso).

The Colombian experience presents a case where the decentralised and small-scale production of coffee and the vast institutional architecture supporting farmers resulted in exemplary institutional and development results. The violence that has for decades affected Colombia has been lower in coffee growing areas, mostly due to the existence of strong social and service networks, associative institutions and the steady flow of income the crop has provided. The Colombian Government, the FNC, national civil and civic actors and international development partners have encouraged coffee production – and the strong institutional capital embedded in it – as a viable and sustainable channel to reintroduce rural populations to lawful and meaningful productive and economic activities (USAID, 2010; ACDI/VOCA, 2014).

Better prices for labour-intensive commodities translate into improved rural incomes and higher opportunity cost of joining armed activities. Dube and Vargas (2006, 2007) found that the fall in international coffee prices that affected the sector in the 1990s led to a disproportionate rise in conflict in coffee producing areas. Their results suggest that, for labour-intensive agricultural activities, income is critical in determining how price shocks will affect insurgency. This explains why international development agencies and the Colombian government have traditionally supported schemes encouraging coffee production in new and abandoned areas. They have also focused on increasing productivity in already producing plots as a way of advancing the peace-building agenda and processes. These findings have supported the implementation of development assistance programmes. For example, USAID has allocated targeted funds to boost the production of specialty coffee as part of its initiatives to eradicate illicit crops from Colombia's countryside.

In 1999, the United States began an assistance programme with the government of Colombia to eradicate illicit production, drug trafficking and guerrilla activities. The so-called Plan Colombia focused on three goals: law enforcement and penalisation of illegal activities; opening up viable and licit legitimate economic activities in rural areas; and strengthening local and national governance. In 2005, USAID began financing a Municipal-level Alternative Development (ADAM) programme implemented in 106,830 ha of land in areas with demonstrated economic potential. For five years, ADAM encouraged farmers to engage in sustainable and licit activities, mainly coffee, cocoa, and rubber cultivation and cattle farming (Areas for Municipal Level Alternative Development Program, n.d.).

From 2002–2007, separate USAID activities provided additional assistance to the specialty coffee sector but always as part of the efforts to open up viable and sustainable legal economic activities in rural areas. After its initial stage, the project was renewed. Its second phase has sought to consolidate the physical, managerial, marketing and human capital infrastructure that the farmers require to produce, sell and benefit from the production of specialty coffee. During the 2007–2012 period, 30,583 Colombian coffee growers were trained in best agricultural practices, staff members at 175 buying points learned about the handling of specialty coffee, and 14 strategic alliances were forged to fasten the commercialisation of value added products (ACDI/VOCA, 2014). One of those strategic social pacts was established in collaboration with Nespresso.

The International Coffee Agreement, the quota system and the crisis that followed

By the beginning of the twentieth century, coffee's prominence as a key global commodity and price volatility was such that producing and consuming countries sought ways to control, influence or intervene in its value chain. In 1906, through the Taubaté Agreement, Brazil staged a unilateral intervention to controlling supply and prices. On 28 November 1940, the USA signed the Inter-American Coffee Agreement in which it set out quotas for coffee producing Latin-American countries. The price floor it established from 1942 to 1946 ultimately resulted in its revocation.

Between 1950 and 1956, climate-related phenomena in Brazil led to a spectacular alteration of global coffee prices. It became imperative to equilibrate demand and supply (Junguito and Pizano, 1993 in Posada and Pontón, 2000). In October 1957, Mexico, Brazil, Colombia, Costa Rica, El Salvador and Guatemala signed the Mexico City Agreement by which they agreed to limit exports. This agreement was never honoured. One year later, 15 Latin-American countries gathered to sign a regional coffee agreement that set the basis for the 1959 International Coffee Agreement. This later effort was joined by a larger number of countries and set the immediate background for the 1962 International Coffee Agreement (ICO, n.d.). During the 1959–1962 period, producers reached different international agreements to allocate quotas to member countries and control prices. Prices, nevertheless, continued to deteriorate (Posada and Pontón, 2000).

The International Coffee Agreement, signed in 1962, structured international commercial activities and controlled prices in the coffee sector for 27 years, until it was dismantled in 1989. The constant fluctuation in coffee prices, and the consequential short-term and destabilising effects, urged 20 exporting and importing countries to form a Coffee Study Group in 1958. As group membership increased, it soon became evident that a global agreement was the only way in which oversupply problems could be addressed. In 1962, the UN Secretary-General called for a conference that ultimately resulted in a five-year International Coffee Agreement (ICA). The objective was to stabilise coffee prices and attain a long-term equilibrium between supply and demand. Until 1974, this measure succeeded in maintaining similar export levels (Posada and Pontón, 2000).

Initially, the 1962 ICA was only concerned with coffee as a raw material. However, at the time the first agreement was still valid, Brazil, the ICO producing member country with the most votes, developed a strong instant coffee industry. It promptly started exporting to the US, the ICO importing country with the most votes. This turned into a dispute upon which the entire coffee international co-operation architecture depended. Subsequent renovations of the agreement included instant coffee as well as part of the allocated quotas for producing countries.

Fully implemented from 31 December 1963, the ICA represented 99.8 per cent of the world's coffee exports from 42 producers, all of them developing countries, and 96.2 per cent of imports in 25 nations, all of them industrialised societies. The accord reduced the coffee world to two sharply differentiated groups. Importers were affluent countries. Producers and exporters were agrarian, developing economies. When this first agreement expired in 1968, annual coffee transactions totalled more than USD 2 billion, employed 20 million people in 50 countries and provided foreign exchange income receipts of 40 per cent or more to six countries in Latin America and five in Africa. A few years later, exporting countries put forward a proposal to update fixed coffee prices and adjust them according to global inflation. This proposition was immediately vetoed by both the United States and Canada (Posada and Pontón, 2000).

For the next three years, during 1972 and 1975, producing countries constituted the sole members of international coffee institutions. By 1976 this situation became unsustainable and a new agreement between exporters and importers came into force. In the late 1970s, Colombia managed to negotiate an increase in its export quotas, marketing over 11 million bags in 1979 and 1980. Subsequent efforts to maintain this level were unfruitful and volumes fell to 9 million bags. At that time, world supply rose sharply and prices remained low until 1985. One year later, prices soared and quotas were temporarily suspended.

Colombia seized the opportunity and exported over 11.3 million bags in 1986, the year in which real prices began to fall until they reached their lowest level in 1993. Throughout those years, the Colombian producers sought to compensate low prices with higher exports. After many years of breached agreements and problems with the allocation of quotas, double markets and inventories, the system collapsed in 1989 (Posada and Pontón, 2000). From then onwards, the international coffee market has been operating in a free manner.

The dismantling of the international coffee architecture was the result of a combination of factors, notably the unwillingness of ICA members to negotiate new quotas (Muradian and Pelupessy, 2005), the rigidity of producing and consumer countries to respond to and control these new coffee allotments (Pelupessy, 2001 in Muradian and Pelupessy, 2005), difficulties in controlling non-members' exports (Akiyama 2001) and the lobbying activities of American firms (Achenson-Brown, 2003, in Muradian and Pelupessy, 2005).

According to Néstor Osorio, former Executive Director of the International Coffee Organisation, crumbling coffee prices represented the worst coffee crisis ever seen in terms of growers' incomes. Discounted for inflation, real coffee prices were among the lowest in history reaching levels that could not support small-scale production, let alone provide adequate income levels for producing households to meet their basic needs. The crisis sent an alarm signal to non-profit certifying and labelling organisations, NGOs and large corporate actors.

To some, the post-ICA scenario led to better price transmission between exporters and producers, a more efficient allocation of resources and the removal of market distortions (Muradian and Pelupessy, 2005). To others, the collapse of the quota system marked a shift in governance power. Far from reallocating resources more efficiently, the end of the ICA meant that farmers in producing countries captured a smaller share of the coffee income (Daviron and Ponte, 2005; Talbot, 2004; and Ponte, 2002 in Auld, 2010). The ICA represented an inter-governmental platform. When this forum was dismantled, the roasting sector began dictating the terms of international trade, which pushed farmers into survival strategies: migration and the cultivation of alternative agricultural produce (Perez-Aleman and Sandilands, 2008). The sudden liberalisation of the industry cascaded into rapid and dramatic repercussions on the livelihoods of millions of coffee growers around the world.

In the absence of proper and timely adaptation, mitigation or competitiveness enhancing measures in almost all major coffee producing countries, over 100 million coffee growers, processors, traders and retailers around the world were affected (Fairtrade International, 2011). Regretfully, short-term policies and actions remain the norm. Discussed in detail in a previous section, the current coffee rust disease outbreak has, once again, exposed how reactive and ill prepared the industry is to respond to large-scale threats to supplies and livelihoods (Mandell, 2013).

Small-scale farm families initially reacted to their crashing household incomes through a combination of increased migration and declining expenditures in education, health care and housing (Varangis et al., 2003 in Bacon et al., 2008). Negative impacts rippled through coffee dependent economies. In Central America, the crisis reached dismal proportions. The World Food Programme declared a food security emergency in coffee producing regions. The Salvadorian government told its farmers not to harvest their coffee because the price they would receive for their crop would not cover their costs (Henríquez, 2003 in Linton, 2005). In Guatemala, aid agencies referred to the situation as a national emergency (Garmendia, 2003 in Linton, 2005). Unsurprisingly, it was precisely in these slow-reacting producing countries that farmers were most dramatically pulled down below poverty lines (ECLAC, 2002). Macro and micro economically, countries suffered. In Colombia, for example, the economic contribution of coffee to the country's GDP decreased from 6 per cent to 2 per cent between 1970 and 2003 (Cárdenas and Junguito, 2009).

The coffee price crisis that started in the early 1990s and that deepened between 1999 and 2003 had enormous social and economic impacts on coffee producers. The consequences are still felt, even when prices have rebounded to previous levels. Prices might have recovered but the coffee crisis persists and farmers continue to live in poverty (Giovannucci and Koekoek, 2003). It was precisely in response to declining terms of trade, crippled rural economies and disposed farmers that new governance structures emerged in the sector. The end of the ICA resulted in the development of several self-regulatory systems in the form of verification, certifications, suppliers' codes of conduct and sustainability-oriented initiatives.

As coffee prices plummeted, a series of stakeholders in the sector intensified their efforts to expand the number of farmers taking part in certification and verification programmes (Bacon et al., 2008). It was also at this time that global markets for certified coffee beans burgeoned. Great expectations have been placed on such efforts. In a way they have become a panacea for all sustainability, productivity, environmental and social challenges faced by coffee growing communities. This enthusiasm for sustainability-oriented programmes has rallied wholehearted support and fierce criticism, both pointing at the need to dispassionately determine what standards can realistically achieve. This book will look in detail at the implementation of two private schemes pushing for more sustainable, productive and profitable coffee endeavours, namely the AAA Nespresso Program and the Nescafé Plan, in Colombia. These two examples of Nestlé's business model to 'Create Shared Value' are discussed at length in subsequent sections.

Half full cups

International coffee prices have historically been characterised by long periods of low prices followed by short periods of high prices. When negotiations to renew the 1989 International Coffee Trade Agreement were suspended, coffee prices fell to less than USD 0.80 per pound. Five years later, a new agreement was reached but as global prices or quantities were no longer controlled, when frost threatened Brazilian harvests, the price of coffee temporarily skyrocketed to USD 2.80/lb. In October 2011, prices reached a 30-year low. A dramatic fall to USD 0.45 per pound triggered a coffee crisis and created havoc for numerous coffee farmers, the majority of them smallholders. Market prices were insufficient to cover production costs, let alone to provide the income coffee growing families and communities needed to survive (FLO, 2011). In Colombia, rural economies were particularly hard hit since slightly over 61 per cent of the coffee farmers allocate less than 1 ha of land to its production, and nearly 90 per cent of them produce coffee on less than 3 ha of land.

Since the end of the international quota system and the subsequent collapse of world coffee prices in 1989, the importance of coffee to Colombia's economy has been declining steadily. The country was unable to adapt expediently enough to the deregulated, and open to competition, new international coffee regime. The sector became vulnerable, partially stagnated and began experiencing a steady decline. During the last two decades, there has been a reversal of fortunes. Coffee production in Colombia has become uncompetitive and unable to firmly position itself in the international market.

Between 1989 and 2011, the country lost 7 percentage points of its market share in global coffee production, contributing with only 7.6 per cent of the world's output between 2000 and 2011 (Fairtrade Foundation, 2012). Colombia once accounted for 13.5 per cent of the world's coffee production. This made the crop an important determinant for the total and agricultural gross domestic product and foreign currency earnings. The importance of coffee for the

Colombian economy had now declined to around 0.61 per cent of GDP and 6.46 per cent of agricultural GDP by 2011.

For three years, the international prices for Colombian coffee, in constant 2010 dollar values, declined to an all time low, below USD 1.00/lb. This poses a serious economic challenge to all coffee farmers, the vast majority of whom are smallholders owning plots of 1 ha of land or less. Many of the growers have not even been able to cover their costs of production in recent years. This challenging scenario, along with tighter supplies and higher price volatility, has put the whole domestic coffee sector, and the many families that depend on it for their livelihoods, at considerable risk.

In fact, from 25 February to 8 March 2013 coffee producers throughout Colombia organised a nationwide strike demanding more economic support from the central government. Subsidies have been granted and the structural reforms that could have fostered long-term development in the coffee sector in particular, and agriculture in general, have been pushed to an undefined future date. A few weeks later, on 18 April, a new agrarian and popular strike was staged. Partly organised by the *Cumbre Nacional Agraria: Campesina Étnica y Popular* (the Peasant, Ethnic and Popular Agrarian National Summit), one of the aims of this new popular movement is to renegotiate farmers' debt owned to the Agrarian Bank. Its agenda goes beyond agriculture. It calls for the creation of a fund to strengthen farmers' and ethnic economies and the evaluation of a United Nations-supported commission verifying the respect of human rights (*El Tiempo*, 2014).

In response to these demands, the government announced that 85,000 farmers with past-due loans for up to COP 20 million would receive COP 300 million in interest-free financial assistance to liquidate their debt. Granted subsidies and financial support have been popularly welcomed but considered insufficient. On 27 June 2013 and 2014 – the Colombian Coffee Day – farmers found they had few reasons to celebrate. The national holiday was declared to acknowledge and commemorate coffee's importance for the domestic economy and for millions of families, the role coffee growers play, as well as the qualities of Colombian coffee that are recognised worldwide. In reality, Law 1337, proclaimed in 2009 to entrench coffee's economic, social and cultural importance in the country, has meant little to farmers. Instead of celebrations, they staged demonstrations demanding more pragmatic solutions from the government. Farmers demanded government institutions looked into the possibility that producers' organisations, co-operatives and guilds made bulk input purchases, mainly for fertilisers and pesticides. Fewer intermediaries would bring prices down, reduce production costs and partially alleviate the economic burden that most farmers in Colombia face (Domínguez, 2014).

Coffee production and international prices have gradually recovered in the last 18 months. It was expected these somehow better conditions would put an end to demonstrations and discontent. However, to September 2014, only partial agreements between farmers and the government had been reached. Mobilisations have been periodically staged and roadblocks recurrently built.

Social unrest has also caused interruptions in humanitarian operations and limited the access to basic services in various regions (OCHA, 2014). It seems, this time remedial measures will not be enough. It may be time for the government to adopt more holistic, more comprehensive and more systemic and structural measures. The National Policy for Rural Development offers such an opportunity.

The promises of a national policy for rural development

Rural societies have traditionally been at the centre of the origins of prosperity. Coffee is Colombia's quintessential example. At the same time, decades of underinvestment and neglect have also put rural communities at the centre of social upheaval. Unfortunately, the economic, social and environmental muscle rural areas exercise has been taken for granted. Despite the key role agriculture plays in most countries and for most societies, the majority of coffee growers, and farmers in general, have wearily assumed agriculture is only a means to subsistence. This attitude is a symptom and reflection of the structural limitations, and new risk factors, that curtail the viability of coffee production as a profitable activity.

The *Paro Cafetero* that went on from 25 February to 8 March 2013 and that has been followed by subsequent popular mobilisations by coffee growers in 2013 and 2014 is a result of this. Other producers have also joined the demonstrations. The demands are shared: to reverse the state of disregard that keeps agriculture as a low-profile sector and business. This is a task that, given its magnitude and pervasiveness, they cannot undertake alone. Complex and deeply rooted difficulties call for institutional support, responsive and inclusive national policies and scaling up successful practices. It is difficult to imagine how subsistence-level, small producers in the middle of Colombia's highlands could possibly be able to tackle, on their own, the many fronts that affect, threaten, influence and shape coffee production.

Sustainable agriculture production processes, comprehensive supply chains and rural prosperity will not be achieved unless efforts are institutionalised, scaled-up and integrated into new or existing policies and practices assembled into self-sustaining models. Policies, however, have tended to see the rural sector as detached from the country's overall development process and growth path (De Ferranti et al., 2005; Parra-Peña, 2013).

Rural–urban articulation and convergence has been overlooked. Growth-oriented policies have tended to focus on cities, neglecting the considerable contributions rural areas have always made to overall national welfare, employment, food production and service delivery. Aggregated gains, from and for both sectors, could be even greater if rural incomes and livelihoods grew and came from a diversified and endogenous productive base. This is without doubt an essential element to ignite rural development, although not the only one. Agricultural activities are far from being the only motor of rural growth. Assets and capabilities in rural areas are plentiful and so is the potential to seize those endogenous resources to promote growth in non-agricultural, yet rural productive sectors.

President Santos alluded to this rural and agricultural promotion policy in the Foreword. This task presents an enviable and not to be missed opportunity to rally domestic and international political, institutional and financial resources to spearhead veritably inclusive and comprehensive growth in Colombia. It could also provide a platform to channel public, private and social efforts and direct them at shifting agriculture away from subsistence and into profitability and sustainability. One of the most effective ways to reduce poverty is to raise investment in the agricultural sector and allocate those resources to the activities and groups that can generate inclusive and endogenous growth. As a spillover, rural expenditure can create new opportunities beyond agriculture (Hebebrand, 2011).

In a World Bank study, De Ferranti and a team of researchers (2005) warn against the inefficiencies that have tended to characterise rural public expenditure in Latin America, Colombia included. Rural support has mostly taken the form of targeted subsidies, favouring specific groups, whilst failing to build a more robust infrastructural and service platform for the sector to grow and develop. Extending the provision and availability of public goods and services in rural areas is imperative to ignite, support and seize positive changes for the millions who produce foodstuffs and deliver services for the rest of the country.

The private and development sector has increasingly built clusters around key productive activities. Nestlé's role in Colombia's coffee sector is such an example. Individual farmers, coffee growing communities, coffee clusters and coffee production as a whole have benefited as a result. In fact, any investment in infrastructure, public goods and services, including education – formal or technical – can generate sizeable societal, economic and environmental gains. But as long as these efforts are not articulated into a coherent policy, the potential for rural progress will neither be fully realised nor permanent. The government is ultimately responsible for channelling, framing and leading the efforts necessary to scale-up sustainable coffee production at the national level (UNDP, 2011).

Responsibility does not mean acting alone. The government can expect to be greatly assisted in its efforts to revitalise Colombia's coffee sector by a plethora of public, private, social, civil and development actors (UNDP, 2014). Together, all these stakeholder and shareholder actors have shown the willingness and commitment to form the kind of inclusive platforms that can build sectoral capacity and tackle structural factors to overcome production risks. The Nescafé Plan and AAA Sustainable Quality™ Program provide two positive examples. These initiatives have integrated market realities with sustainability issues and used sustainability as a marker for present and future brand competitiveness. As such, they could certainly be used as starting points upon which to build to make the trade more resilient, competitive, profitable and dignified.

★★★

Like millions of coffee growers all over the world, the Colombian coffee farmers are facing mounting pressure to better manage increasingly scarce and more expensive inputs; overcome the competitive perils from poor and absent physical, financial, educational and associative infrastructure; cover high internal transportation costs; and acquire and put into practice relevant skills and knowledge to grow good quality crops more effectively and efficiently. On top of this, increased climate variability – prolonged drought, higher temperatures and altered rainfall patterns – make the harvest seasons even more unpredictable, raise production costs and reduce growers' income.

Profitability has been further eroded by the sharp rise in the cost of fertilisers and also due to labour shortages. Even in oil and gas rich Colombia, the prices of chemical inputs may represent up to one-third of coffee production costs. All these factors are contributing to producing and sourcing high quality coffee a steadily more difficult and complex process. They also point to the pressing need to restructure and revitalise the coffee sector so that it produces better quality coffee. Also better paid, these higher quality coffee beans are the result of the implementation of the sort of socially, economically and environmentally sustainable practices that allow farmers to lead better lives.

This chapter was concerned with the many difficulties farmers face as part of the supply chains that connects them with coffee drinkers all around the world. In addition to sector-related difficulties, women growers still have a few extra battles to fight. They have to face gender discrimination in the public and private realm that limits access to the resources and knowledge that can bring prosperity to them, their farms and their families. If this were not enough, pendulum-like climate variability and fluctuations are adding yet another wild card to an already highly volatile sector. These elements are adding up to an ever more complex scenario that is putting at risk the livelihoods of the two million people that depend on this crop in Colombia alone.

This complexity suggests that the world's largest coffee purveyor, and other equally big and small companies all the same, may have to worry about their future ability to procure the amount and quality of beans they need. Retailers, roasters and traders are all too aware that demands in the mainstream and specialty markets in both producing and consuming communities are escalating. They also know that poverty, low productivity, limited market access, low levels of human capital accumulation and credit constraints are some of the sustainability issues affecting the security of coffee supplies.

These are no small reasons to take action and transform the coffee landscape in one of the most important coffee producing countries in the world. But if current trends hold, sustainability-oriented practices will become a differentiator marker, separating one brand's coffee products from the many other competing goods in the instant and single-serving coffee markets.

Note

1   1 USD = 2,017.82 COP to 17 February 2014 (Bloomberg, 2014).

4 Nestlé in Colombia

Founded in 1866 in Vevey, Switzerland, Nestlé became part of the Colombian food and beverage industry in 1922 when the first imported products, mostly confectionery, began to arrive in the country via Panama. In the following 90 years, the company would grow larger in scope, size and operations. It currently offers Colombian consumers a wide array of both innovative and staple products meeting, at every life stage, a wide range of food and nutritional needs, demands and desires. It has an extensive and robust industrial presence that includes five factories, one main national distribution centre and six cross-docking platforms. Its administrative headquarters are located in Bogotá.

Nestlé Colombia has been operating in Bugalagrande (Valle del Cauca) since 1944, producing culinary and dairy products, as well as coffee and beverages; from 1971 in Dosquebradas (Risaralda) where confectionery, cookies, chocolates and wrappers are manufactured; pre-condensed milk in the Florencia (Caquetá) factory, built in 1974; in Mosquera (Cundinamarca) making pet food since 2002; and processing dairy products in Valledupar (Cesar), as a result of the establishment of a partnership with DPA (Dairy Partners Americas) in 2004 (Nestlé Colombia, 2013a). Around 90 per cent of the portfolio available in the country is manufactured locally. Its products can be found in over 300,000 sales points throughout Colombia and in eight out of every ten households in the country. This grounds the company's procuring, manufacturing, production and distribution activities firmly within the domestic economic landscape.

In September 2013, the Swiss company was supporting more than 8,800 individual suppliers and its industrial architecture directly employed 2,362 workers and indirectly some additional 1,738 people. It was previously discussed how employment is indeed an important way to reduce poverty, especially when jobs are formal, stable, socially recognised and profitable. However, Nestlé's overall impacts on Colombian society extend far beyond its workers and consumers. Through its supply networks, the company is making a positive difference to farmers' lives via quality and productivity improvements in the sourcing communities from where it procures raw materials.

Connected through supply chain steps, farmers, suppliers, factories, employees, consumers and national and international shareholders are tightly and interdependently related with one another. By making coffee production a more attractive, profitable, dignified, competitive and professional economic activity, Nestlé is contributing to the wellbeing of almost two million people who depend on the crop for their livelihoods in one way or another. The challenge and merit of Nestlé's approaches reside in consciously and deliberately bringing about valuable benefits and prosperity to a large number of stakeholders, simultaneously and through mutually reinforcing activities.

The efforts Nestlé Colombia has made over time to meet consumer expectations, deliver high quality products, follow fair business practices, offer a good work environment, be socially committed and yield strong financial results have been widely recognised. The various Nestlé factories in Colombia have invigorated local economies; become pivotal local economic actors; human-capital enhancing employers; quality referents; and more recently, solid community stakeholders, shareholders, mobilisers, innovators and prosperity facilitators. This corporate reputation and business model have positioned the company as the largest coffee buyer in the country and has helped to consolidate and expand its market penetration and shares. In 2011, the Reputation Institute and Goodwill Communications ranked Nestlé Colombia, the company's fourth largest market in Latin America, as the number one most reputable firm in the country (ANDI, 2010; Dinero, 2011).

To coffee farmers, producers and suppliers, Nestlé aims to stand out as a predictable, stable, reliable, trustworthy, quality-oriented and committed long-term partner. To its consumers, the brand seeks to equate to consistent good quality. To its workers, the firm is working to assure a safe and enriching corporate environment to grow professionally and personally. To its shareholders, the company is pursuing a model of consistent and organic growth and profitability. Put together, these pieces are part of Nestlé's 'Creating Shared Value' approach to business.

Sharing value locally

For Nestlé Colombia at large and certainly for the coffee-producing Bugalagrande plant, which produced 34,837 tonnes of finished goods in 2012, reaching consumers all across the country can only be achieved through an extensive distribution network, especially in the countryside, where road infrastructure is poor. It is an equally challenging task to get Nestlé products strategically and visibly shelved in family-kept shops and independent supermarkets. The company is thus aiming at widening its existing extensive distribution channels even further so that its products can be regularly and efficiently distributed to consumers living in the most remote parts of the country.

In 2007, the company devised a training programme, macro-allies (in Spanish '*macroaliados*'), directed at teaching freight entrepreneurs how to develop a formal business plan, identify sales potential in a given geographical area and make their businesses more efficient and competitive. Nestlé was in fact looking at ways of getting more of its products to reach more places, more consumers, in less time and generate further demand. During the seven years the programme has been implemented, the scheme has proven successful: 61 distribution firms formally employ more than 870 sellers, and have yearly individual accounts for more than COP 3,000 million, the equivalent of almost USD 1.5 million[1] (Nestlé Colombia, 2013b). Beneficiaries are now better prepared in terms of human capital development, leadership, accounting, stocking, capacity building for sales teams, etc.

Whilst the initiative proved greatly beneficial to participants, for Nestlé, current processes to get a product from the factory line to the shelves are shorter than they ever were in the past. Already-established rural distribution networks are being harnessed to expand the array of goods supplied to a community. They are also being used to identify the specific local needs of potential customers belonging to different socio-economic groups at the base of the consumer pyramid. New distributors, for their part, have gained new planning and management tools to organise their business ideas and put them quickly into practice (Nestlé Colombia, 2013a). In addition, a two-track distribution approach was put in place to get Nestlé's products to all of its actual and potential customers. It specifically targeted small corner shop owners and independent supermarkets.

Once the goods had been taken to lorries for distribution to corner stores, Nestlé needed to strategically and visibly display them. In 2010, the company began training small shop owners on basic business entrepreneurship. Since then, the programme has bolstered loyalty among 38,000 shop owners. Working with many different consumer and economic groups at the base of the income pyramid has widened the company's distribution network as well as making it more effective. This distribution channel currently represents half of the total sales of Nestlé in Colombia. Collaborating with small shop owners and their families also exposed the many limitations they face in terms of access to basic services such as education, health and dental care. These needs were soon transformed into an opportunity for product and market innovation.

With an initial three-year investment of USD 2 million, the Nestlé Wellness Shop (*Tienda del Bienestar*, TBN, in Spanish) scheme was created to help families gain access to dental and life insurance in case of untimely death of the heads of the households. Up to July 2013, some 5,000 households have benefited from this programme (Nestlé Colombia, 2013b). In 2012, this initiative received the VENN award in the Consumption Category, a recognition extended every two years by the National Business Association of Colombia (ANDI) to highlight and promote good commercial practices at the selling points and to the benefit of the final consumers (ANDI, 2010).

By 2012, the company established a partnership with Bogota's Jorge Tadeo Lozano University to design the Super Allies (*Super Aliados*) initiative, aimed at widening distribution channels with independent supermarkets. That year, an initial investment of COP 300 million financed training sessions for more than 100 people in leadership, accounting, finance, logistics and marketing. Independent grocery stores gain the professional knowledge that can help them become more profitable and competitive. Nestlé thus contributes to keeping small and larger shops well structured and, in return, can enhance and multiply product coverage in all of those selling points.

The same year the Super Allies programme was set into motion, the company was selected as a 'Success Supplier' (*El Proveedor de Éxito*) by the ÉXITO supermarket chain. Created by the Éxito Group, the Successful Supplier Contest began in 2009 as a way to recognise the conglomerate's commercial partners (Grupo Éxito, 2013). The award recognised Nestlé Colombia's dynamism, innovation and role as an 'exemplary supplier in logistics practices' (Grupo Éxito, 2012). In addition, the Éxito Group emphasised the assistance in the development of business strategies it has received as a result of the close collaborative relationship that exists with Nestlé Colombia.

These initiatives are examples of how Nestlé is responding to its social and economic context in a way that is profitable for the firm and beneficial to a wide array of stakeholders. The company needed to widen and extend its distribution network in a country where road infrastructure is poor and a large share of products is purchased from corner shops. It thus went on to train freight entrepreneurs in business management, sales, accounting and stocking. It also started working with small shop owners on basic entrepreneurship models. As a result, Nestlé products are being transported to 38,000 shops in the most remote parts of the country by an emerging group of businessmen. Other positive externalities from enhancing product coverage include the development of business and entrepreneurial skills and capacities amongst small and independent grocery stores. Nestlé has gained more selling points, grocers a new set of skills that can lead to growing profitability and competitiveness.

Human capital

Nestlé is well aware that its employees are one of the company's pivotal resources to build, consolidate and innovate its competitive advantage. In the coffee processing Bugalagrande factory, the firm has taken proactive and welfare-oriented moves to provide a competitive reward system; a safe and healthy work environment; and meet workers' aspirations with training and continuous development opportunities. According to the Chairman of the Chamber of Commerce of Tuluá, Nestlé has set the standards for training, capacity development and human capital formation for the region (personal communication).

The Bugalagrande factory has historically paid the best salaries of the region for similar types of work, which are well above inflationary levels and the legal minimum wage. Salaries, responsibilities, expectations and benefits are formalised in formal work contracts. In a country where informal unemployment averages 62 per cent (Banco de la República in Portafolio, 2012), opportunities to land a job in the formal sector are highly sought after. This is partly due to the promise of predictable and regular salary payments, as well as benefit packages that may or may not be honoured. According to Colombia's labour law, workers are entitled to receive dividends from the enterprises they work for equivalent to one month's salary per year of work. However, in 2012, it was estimated that up to 88 per cent of all workers in Colombia did not receive such dividends, 67 per cent did not get paid holidays and 62 per cent made no contributions to the health and pension systems (Portafolio, 2012). Nestlé, on the other hand, is known for offering competitive employment packages.

From 1990–2013, the wage ratio between Nestlé and the government's minimum wage averaged 2.51. Nestlé records from 1990 show that the conventional minimum wage paid in the Bugalagrande factory has been consistently higher than the minimum legal requirement. On 31 August 2013, the factory employed 658 workers under indefinite term contracts and a further 176 under fixed term conditions. Operating the coffee division alone provides work for 126 people, 110 hired through indefinite term contracts and 16 more for fixed periods of time or to perform specific tasks (Nestlé Colombia, 2013b).

In February 2013, the Colombian Government increased the monthly minimum wage by 4.02 per cent, from COP 566,700 to COP 589,500 (Decree 2738; Ministerio del Trabajo, 2013). In comparison, the base salary for Nestlé workers increased from COP 1,077,816 in 2012 to COP 1,099,380 in 2013, a remuneration 86 per cent higher than the then official level. Throughout the years, wage differentials have gone from an all time high of 182 per cent in 1999 and 2000 to 86 per cent in 2013 (Figure 4.1). On average, salaries received

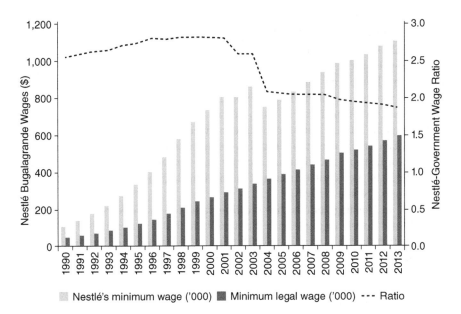

Figure 4.1 Nestlé Bugalagrande–government wage differentials
Source: Nestlé records

by workers at the Bugalagrande factory have been 150 per cent above those stipulated by the government (Nestlé records; Ministerio del Trabajo, 2013).

Higher wages are only one way in which the company is able to attract, retain and harness some of the best available human resources at all skill levels in the region. In 2007, the average number of years of service of direct Nestlé workers was 15 years, which already confirmed the level of employee satisfaction and the opportunities to start and build a career in the company (Silverman, 2007). It also illustrates the existence of corporate channels available to all tiers of workers to acquire new skills and hence more knowledge, greater responsibilities and better economic prospects. The more skills employees have, the more proactive and strategic role they play in the company's long-term growth and development.

Investments in better health, education, nutrition and housing, and also higher incomes have resulted in improved social and development indicators for the workers and their families. On and off work, Nestlé employees expressed their satisfaction with the support the company offers on three pillars: nutrition, health and safety. Within the factory premises, the company is committed to maintaining high operational standards and attaining zero injuries and illnesses in the workplace. At the same time, contracts were negotiated with the workers' unions Sintraimagra and Sinaltrainal to provide health benefits to all workers and their direct dependants. These include free medical consultations with the factory's doctors, eye care support and subsidies for certain medicines, dental services and orthopaedic expenses of up to 70 per cent. The National Workers' Union of the Agri-Food System, Sinaltrainal, represents some 3,100 workers. Founded in 1968, Sintraimagra is the National Trade Union of Workers of the Food and Fat Products Industry.

Since 2001, the company has made 720 loans worth COP 7,605,635,447 to its employees at the Bugalagrande factory. Credit is usually used as mortgages, to cushion household shocks and to make long-term capital investments. In 2012, 38 worker-members of this revolving fund received COP 1,654,272,165 in loans at 0 per cent interest. Commercial banks do offer similar loans but at much higher interest rates, generally between 11.22 and 13.54 per cent. Nestlé's favourable credit facilities have contributed to the encouragement of asset accumulation and housing security and upgrading among workers. To June 2013, 92 per cent of the staff at the Bugalagrande factory owned their own houses (Nestlé records).

Nestlé in Bugalagrande is thus creating value for its workers as it explicitly, broadly and deeply engages with the household and communities they belong to. Staff members are seen as part of a family and a community and thus the company has extended benefits to dependents and facilitated the implementation of collective, academic, civic, wellness and sports initiatives that help build social capital. For instance, and in accordance with Nestlé's aim to enhance the nutritional and wellness status of its first tier of consumers, factory products are available to workers, and the community in Bugalagrande, at discounted prices. Employees are also entitled to a monthly milk stipend.

The company supports new parents with yearly subventions and a one-time delivery of 35 cans of baby formula; provides income support and scholarships for workers' children to attend school and pursue undergraduate studies; distributes school supplies and uniforms; and gives a transportation allowance for children to commute to the academic institutions they attend. For over 40 years, the firm has contributed to a sports fund and sustained a committee organising a wide range of sports, recreation and cultural activities. In this way, the company takes care of its employees' children, directly contributing to their educational needs as well as their wellbeing. These actions also enlarge their cultural, professional and nutritional opportunities.

In June 2013, 50 per cent of the children of the staff had benefited from scholarships awarded by the company and had taken full advantage of 27 apprenticeships at the Bugalagrande factory that were available to workers' dependants. This policy ensures that workforce is secured and then retained from the nearby communities, increases the pool of locally available talent and human resources and lays down the foundations for a more robust and better-prepared recruiting base. Since the factory opened in 1944, two to three generations of workers have often been linked to the company. More importantly, many of the second-generation factory workers and community leaders have been the direct beneficiaries of the company's scholarships and educational schemes.

Continuous improvement of human capital has ultimately translated into better earnings, progressive increments in quality of life, higher purchasing power and shared focus on life–work balance and wellness. Collectively these actions offer safe, competitive and encouraging working and living environments for the factory's staff and their families. In fact, some of the interviewed employees stated that one of the things that drew them to work in Nestlé Bugalagrande is the fact that the firm does not see workers in isolation but as part of a family unit (group interviews with factory staff, May 2013). The synergistic impacts of these factors contribute to the steady strengthening of Nestlé's long-term competitiveness. It also helps to forge strong partnerships and relationships with those directly or indirectly linked to the company's activities, namely public, social and civil actors (Nestlé records; Nestlé Colombia, 2013b).

It should be noted that the labour relations between the Nestlé Bugalagrande factory and the two unions that represent the workers could be improved considerably. Confrontations and disagreements between a company and its unions are not unusual. According to Alliance Sud, the Swiss Alliance of Development Organisations, the pervasive and acute environment of distrust that exists between the company and the union to which most of its workers belong to, Sinaltrainal, have in the past become unmanageable. The escalation of disagreement was at some point such that it was proving to be counterproductive for both sides, as well as for the employees and the surrounding community (Egger, 2011).

Sinaltrainal was formed in Bugalagrande, which soon made the municipality a strong hub for the workers' union. However, the relationship between the union and Nestlé deteriorated quite significantly during the 1980s and their positions became antagonistic. Since 1986, 13 employees and ex-employees have been assassinated without any convictions, and most of the victims have been Sinaltrainal trade union leaders. The union claims that Nestlé is somehow, and at least indirectly, responsible for these deaths.

It should be noted that independent observers like the government in Bogotá and the International Labour Organisation (ILO) have both refuted these accusations. Moreover, it seems that the generalised level of violence unions had for long witnessed in the country fell considerably since Angelino Garzón, a former trade unionist and Colombian Ambassador to the ILO, became the country's Vice-President (Egger, 2011). Despite the tensions that exist between Nestlé and Sinaltrainal, virtually all workers at the Bugalagrande factory are unionised, a ratio that is many times higher than the national average of 4.7 per cent (Egger, 2011; Mora and Tovar, 2011).

As the biggest employer in that area, Nestlé has created its own multiplier effect in the local economy. This phenomenon has been fuelled by the aggregate demand for services and goods spurring from the disposable income in the hands of factory employees and their families. In a similar fashion, factory-community interdependence implies that circumstances negatively affecting the plant's daily operations will result in unfavourable externalities. According to Bugalagrande's staff, union negotiations create a tense environment in the municipality, partially disrupting commercial and social activities alike. Beyond economic prosperity, strong community-factory relations have proved crucial to normalise daily life at times of conflict or violence. The links between prosperity and stability that have been forged over the years between Nestlé's factories in Colombia and the neighbouring communities are issues turned to in subsequent sections.

Quality archetypes

For Nestlé, product quality is non-negotiable. This emphasis is at the core of the company's vision as a Nutrition, Health and Wellness company. It is also a deeply engrained principle in the company's activities all around the world, and one that has been successfully passed on to its factory workers as the most important rule that they must follow. During a group interview with staff members belonging to different departments of the Bugalagrande factory, they all repeatedly made reference to the emphasis they are required to make on quality control (Nestlé Colombia, 2013b).

The Swiss company is committed to predictable, consistent and constantly improving quality. This corporate value has brought about substantial changes in the mindsets and working practices of its employees, ancillary firms and competitors through so-called demonstration effects and spillovers. Other players in the region are looking at the company and learning from it, taking the firm's benchmarks as reference points for their own commercial activities (personal communication, Chairman of the Chamber of Commerce of Tuluá). The links the corporation holds with local economies and societies induce the transmission of new ideas and more efficient practices. As part of a mutually reinforcing process, local expertise and knowledge are also infused into corporate activities and operations.

Workers are the company's very first tier of consumers and first-hand product advertisers and endorsers. It is crucial that they trust the brand they are part of and are fully confident that the Nestlé goods they find on the shelves are of good quality. Employees informed the study team about the freedom they have to halt any process if they have any concerns about the production process. They know how groceries and edibles are being manufactured, how raw materials are selected, the quality standards suppliers have to comply with and the stringent controls imposed at all stages of production. They are aware that quality controls are uniformly enforced. Whatever is demanded from them in a given department applies to the rest of the factory and across all of the company's plants in Colombia and abroad (personal communication, Nestlé Bugalagrande, June 2013).

Moreover, factory workers recognise the large and widening quality gap between Nestlé and local producers. This positions Nestlé well ahead of its competitors and turns it into a quality reference point for suppliers, ancillary firms, competitors and consumers alike. At the same time, widening quality gaps suggests many of Bugalagrande's local suppliers are lagging behind. In its Nestlé Creating Shared Value and Rural Development Report for 2010, the company touched on the opportunities for rural growth and cluster formation its factories open up (Nestlé SA, 2011). The establishment and operation of a Nestlé plant can potentially act as a pull and cost-reducing factor for smaller firms to settle in its vicinity. Cluster, nevertheless, can only be formed if competitive entrepreneurial skills exist amongst the local business class.

In Bugalagrande, the ample quality and skills gaps between Nestlé and local firms have become a matter of concern for the regional Chamber of Commerce (CCT). Tuluá's Mayor and Director of the Chamber of Commerce agree on the positive, productive and competitive business environment that exists in Bugalagrande. They attribute this growth-conducive entrepreneurial fabric to the operations of the Bugalagrande plant; a trait they consider sets the municipality apart from others in the region. They believe the municipality is a good place to start a business and make commercial transactions. It has better infrastructure and services than other towns. It is also more peaceful and prosperous. This fosters continuous and reliable local development (Chamber of Commerce of Tuluá, personal communication, May 2013).

The factory creates demand for vendors, suppliers and providers of goods and services. Yet, local supply has often fallen short in terms of quality control even when ex-Bugalagrande factory workers have formed small firms delivering a wide array of services to the factory. In fact, differentials in Nestlé's quality and human capital requirements from locally available resources have continued to widen. This is a growing preoccupation for the local Chamber of Commerce and local business associations. They see this lag as detrimental to the municipality's future commercial and agricultural development.

Undoubtedly, the government needs to augment investments in metier, professional and entrepreneurial education as well as in business incubators. An even bigger challenge still is for the government to take the lead in setting the region's development priorities and building a broader and cohesive network of partners. In this way, the company, other private actors, local non-governmental organisations and Bugalagrande's inhabitants can keep an open dialogue, and work together, to make the type of social and economic contributions that can further contribute to the development of the community. Examples from the past offer valuable insights and a course of action that can be at least partially emulated.

When it first opened, the factory faced a shortage of qualified workers. Subsequent actions have since then been taken to identify, build and foster talent and skills amongst employees. For instance, 51 scholarships are granted every year for workers interested in finishing high school, pursue practice-oriented trainings in areas relevant to the factory's activities, attend university, acquire additional skills and capacities or enrol in the courses offered by the National Learning Services (SENA). Additionally, the company finances 50 distance learning education courses and supports one training centre in the facilities of both the unions it works with, Sintraimagra and Sinaltrainal. This is a way to leverage the factory's very own internal human assets, build a culture of unremitting improvement and turn local insights and ingenuity into new business opportunities. The competencies developed improve earning capacity of the workers in and outside Nestlé. Knowledge, know-how, experience and expertise are all transferable human capital assets. Since these are applicable across the labour market, they make human capital gains incremental.

Educational assistance has also been provided in partnership with local institutions deeply engrained in the community and seeking to address important gaps in the entrepreneurial landscape of the region. For 15 years, from 1986 to 2001, Nestlé Colombia SA closely worked with the Sarmiento Palau Foundation to execute a programme for business development in the municipality of Bugalagrande, where the coffee-producing Nestlé factory is located. This initiative supported, trained, built capacities, advised, offered loans and helped commercialise the goods and services produced and delivered by local entrepreneurs. This micro enterprise creation programme assisted over 2,000 business units in the region. Bolstering and fostering the development of new economic activities have continuously created employment opportunities created in the region.

Using its corporate knowledge to the pursuit of social growth, Nestlé has offered training modules on income generation, productivity enhancement and competitiveness, business management, marketing, leadership, human resources management, strategic planning, market research and quality control. Beneficiaries were also visited on-site by expert teams and encouraged them to participate in regional fairs to commercialise their products. These new businesses have done much more than merely improving the income of its owners. These new ventures have become a source of permanent and seasonal employment in the area and have contributed to poverty alleviation. In doing so, they have diversified the economic base of the municipality and strengthened its resilience to external economic shocks (Fundación Sarmiento Palau, 2009; Nestlé SA, 2014a).

Nestlé operates 230 factories in rural areas worldwide. In most of them, and whenever possible, the company has encouraged local farmers to grow crops they can sell back to it as raw materials (Nestlé Colombia, 2013b). Locally produced inputs revitalise rural economies, heighten supplier certainty and contribute to the firm's successful operation. Well aware of the potential economic, reputational and human capital benefits farmers could accrue if they became regular partners to the Bugalagrande plant, the Chamber of Commerce would like the company to support the local production of inputs currently being bought from elsewhere.

Nestlé's Agricultural Extension Services (AES), which have been a defining factor in securing adequate and regular inputs, have been mostly targeted at coffee producers and milk farmers. But since the factory requires raw materials not always available in the vicinity, the firm could benefit from helping local farmers supply them, sustainably increase output and improve the quality of their products. It has been proposed that AES are also directed at encouraging orchard renovation and expansion. This would enhance household food security and provide additional income to many farmers living at or below subsistence level (Chamber of Commerce of Tuluá, personal communication, May 2013). Similarly, it would enlarge the 'community of interests' Nestlé has established in the local community, a key factor in attaining factory and company success and furthering rural development (Biswas et al., 2014). This is a process that, as we have discussed in previous chapters and sections, is severely limited by the absence of roads.

The coffee roads

In 1979, the United States Agency for International Development (USAID) financed and supported the Pick and Shovel (*Pico y Pala*) road infrastructure projects. These initiatives were directed at providing small farmers with better access to local and international markets. The projects immediately improved personal mobility and reduced transportations costs, travel times and efforts. Road upgrading and construction also led to the expansion in commercial banking and government credits. Visits to and from credit agents were more convenient to make, cheaper and faster (USAID, 1982). USAID support for road infrastructure in Colombia in the late 1970s was almost unanimously approved of, especially when the beneficiary populations were involved in route selection and gained employment during the construction works. Forty years later, the state of the roads has not changed. Colombia's still deficient road infrastructure undermines the country's economic competitiveness and productive potential of the agricultural sector because of high transport costs.

Limited and poor road connectivity further exacerbates rural–urban income disparities, amenities, services and development disparities. It also delays the adoption of new ideas and technologies and the delivery of agricultural extension services (USAID, 1982). New and upgraded roads that permit vehicle access will stimulate agricultural production and marketing for producers by boosting local and regional market potential. If isolated, yet productive or potentially arable agricultural land can be connected to main roads, agricultural outputs will increase and so will farmers' incomes as well as access to various services. Early studies have shown that travel times and transport costs could be cut by 80 per cent if rural roads were upgraded in areas previously inaccessible to motorised vehicles. Confronted with limited options to reach outside markets, farmers have intensified the production of goods they can sell locally (particularly perishables) between 50 per cent and 200 per cent (Evans et al., 2006 in Sieber, 2008).

More recently, the authorities of the department of Caldas have resumed a project to build an airport in the city of Palestina. The project was first put forward in 1977. The so-called 'Coffee Airport' was conceived as a way of obtaining competitive advantages that could, in turn, bring about economic and social benefits to the nearby communities. At present, the project is also looking into improving the area's international air connectivity to both passenger and freight transatlantic flights, fostering the profitability of agro-industrial activities (Aeropuerto del Café, 2014).

As a further example of the level of public and private trust deposited on local coffee institutions, the 2002 COL/12034 agreement designated the FNC and the Coffee Growers' Departmental Committee in Caldas as the project's implementing, administering and coordinating authority (Aeropuerto del Café, 2014). The project expects to reduce airfares and transportation risks, generate employment and decrease travel and transportation costs. The benefited population could include over 650,000 inhabitants in the municipalities of Manizales, Neira, Villa María, Palestina, Santa Rosa and Marsella (Chamber of Commerce of Manizales, Caldas, 2013).

The presence of the Bugalagrande factory further stresses the importance of having access to basic service and road infrastructure. During its early years of operation, the plant offered public services to its neighbours, which other people of the region did not have access to. People recall going to the Bugalagrande factory to fetch potable water until widespread water delivery services were available in the region (Durán Castro, Chairman of the Board of the Foundation for the Integral Development of Valle, personal communication, May 2013). The firefighter brigade in the plant was also the first one in the region. Nestlé's presence in the municipality has made it a more peaceful and prosperous place to start a business, and contributes to elements that foster continuous and reliable local development (Township of Bugalagrande-Valle, 2010).

During the 1950s and 1960s, the company engaged in building and upgrading the necessary infrastructure to connect the factory with its suppliers and workers' residences. Dirt roads were enlarged, paved and transformed into main focal traffic arteries. These capital assets opened up many opportunities for further development of the region by facilitating the production of goods and services, the distribution of finished goods, access to basic health and education services, transportation to and from employment hubs, and connectivity with other municipalities. The number of economic and development opportunities in the region significantly increased and improved (Durán Castro, Chairman of the Board of the Foundation for the Integral Development of Valle, personal communication, May 2013). Bugalagrande has better infrastructure and services than other similar size towns. Despite these infrastructural improvements around the factory, to this day the lack of good tertiary roads disproportionately raises costs for farmers and other suppliers of the region (OECD, 2010). Better infrastructure also reflects healthier fiscal balances.

Nestlé is the only medium or large enterprise based in Bugalagrande, and by far the most stable company and important fiscal contributor to the municipality. According to the Bugalagrande factory management, at least one person per family in the municipality is related to the factory and obtains his/her income as a worker, supplier or service provider (Nestlé Colombia, 2013b). A similar key and stable contribution is made to the municipality's tax revenue and budget. As it can be seen in Table 4.1, from 2009 to 2013, the factory's contributions to the municipality's budget have averaged 23 per cent of the total revenues. In those five years alone, tax receipts ranged from a 19 per cent low in 2009 to a 29 per cent high in 2013 (Table 4.1).

The municipality's mayor estimates that 90 per cent of the community's daily activities are related to the factory (Taguado Trochez, personal communication, May 2013). Clearly, some of the most straightforward contributions to the prosperity of host communities Nestlé has made have been in the form of salaries to its employees and suppliers, as well as taxes and other payments to the State at different levels of government.

These fiscal resources are key components of the public revenues that can finance the delivery of social services and address commonly agreed economic, social, environmental and cultural issues and needs of local residents. How these resources are spent cannot be left to politically unaccountable private firms and their corporate social responsibility agendas and initiatives (Friedman

Table 4.1 Contributions of Nestlé to municipal tax income of Bugalagrande, in million COP

Year	Municipality's budget	Municipal tax income	Factory tax payments	Factory budget contributions
2009	15,761	2,561	3,032	19%
2010	15,635	2,868	3,279	21%
2011	16,020	4,033	3,633	23%
2012	15,880	3,847	3,609	23%
2013	13,071	3,881	3,739	29%

Source: Nestlé records

1970; Lantos 2001). It cannot be stressed enough that elected representatives are ultimately accountable for the use of such fiscal contributions and how these affect social welfare. Whilst they are also responsible for formulating the laws and policies that will protect the environment, some private sector firms are putting in place innovative technologies and management systems to reduce their footprint.

Reducing footprints

Food processing factories impact the surrounding environment in terms of energy consumption; water use, treatment and discharge; atmospheric emissions; and solid and liquid wastes management. In 1991, Nestlé articulated its environmental commitments and corporate policy in its Environmental Management System (NEMS). It actually became the company's answer to these commonly shared environmental challenges. Five years later, this tool was taken to Colombia's different factories. Prior to being implemented in the country's factories, NEMS provisions and national environmental laws were compared for compatibility. Where two criteria differed, the highest standards were selected.

NEMS is implemented all across Nestlé's global operations to ensure compliance with environmental policy and to systematically push for the continuous improvement of the company's environmental performance. It also seeks to build trust with consumers, government authorities and other stakeholders and enable the certification of the company's factories to the ISO 14001 international standard.

These seemingly robust sustainability and environment conservation practices are the result of a long corporate process to finding ways of improving the ecological and environmental impacts of the company's operations. Complying with its own internal standards, which are often significantly more stringent than those stipulated by the domestic laws, Nestlé in Colombia invested in a wastewater treatment plant as early as 1987, many years before such facilities were made mandatory by national legislation. At that time, the factory was discharging 22 litres/second of wastewater and 1.5 tonnes of suspended solids (*El Tiempo*, 1992). In 1992, the water treatment plant received the Ecological Merit Prize from the environmental authorities of Valle del Cauca. This facility is considered an industrial benchmark and a guide in environmental education in Colombia (*El Tiempo*, 1992). Every year, over 1,000 students, civil servants and members of the public visit the plant (Nestlé SA, 2005).

These efforts, nevertheless, have not been universally praised. Sinaltrainal union, one of the unions representing Nestlé's workers, has constantly criticised the company for not doing enough to protect the environment and keeping the Bugalagrande River clean. The union has acted as a constant watchdog, urging the firm to be more stringent in complying with local and national environmental laws, its own corporate policies, OECD guidelines and UN Global Compact's principles (Egger, 2011).

In Colombia, non-recyclable solid wastes are disposed of in local landfills. Those that can be reused, cardboard, plastics, oils and tinplates, are delivered to recycling–certified contractors. But these residues represent less than 10 per cent of the total factory waste. In the Bugalagrande factory, cisco, the husk left from the coffee production process, accounts for 90.2 per cent of the generated solid wastes even when the parchment skin represents only about 4 per cent of the weight of a coffee bean (Ernst Basler + Partner, n.d.). Before this coffee by-product could be reused, more than 16,000 tonnes of cisco were being sent to local landfills every year (Nestlé Colombia, 2006). This prompted the company to look into a more environmentally responsible and efficient way of reducing, reusing and recycling this waste.

In 2009, Nestlé invested some USD 12 million improving two key projects: the wastewater treatment plant and the construction of a biomass boiler to convert coffee husk into energy. Borrowing from its successful experience in India, Philippines, Russia, Thailand, and Switzerland and after four years of research, Nestlé constructed the first cisco plant on the continent (Nestlé Colombia, 2013b). Supervised by ten permanent workers, this biomass plant began operations in 2010 and immediately achieved considerable environmental gains. The energy generated substituted otherwise needed natural gas and liquid fuels. This has meant that at least 1,500 tonnes of husk and 80 tonnes of ashes are no longer transported to local landfills (Semana, 2013).

As a result, factory carbon dioxide emissions have fallen and the appearance of coffee-coloured treated wastewaters has improved. Only in terms of emissions, CO_2 release levels from coffee grounds, are 17 times lower than those of natural gas and 24 times cleaner than fuel oil. Coffee grounds emit 3.6 kg CO_2/GJ, whilst fuel oil is responsible for 86.7 kg of CO_2/ GJ, and natural gas produces 61.2 kg CO_2/GJ. From the time coffee is planted to its final incineration in the biomass burner, this energy recuperation process has a neutral carbon footprint (Nestlé Colombia, 2006; Portafolio, 2009).

When the project was completed and fully implemented in 2010, some 4,537 tonnes of coffee grounds were turned into 35,615 GJ of energy. Table 4.2 shows the energy produced from coffee husk from 2010 to 2012. During its first year of operation, the cisco plant produced 31 per cent of the energy used in the Bugalagrande factory. In 2011, 3,105 tonnes of coffee grounds produced 24,379 GJ of energy. This represented 12 per cent of the Bugalagrande facilities' needs, which were 76 per cent higher than on the previous year. In 2012, the factory's total energy requirements amounted to 261,069 GJ, more than double the 2010 levels, out of which 15 per cent were met through the cisco plant.

Every year, the plant's energy contributions vary in absolute percentage terms. Energy generation depends on the volume of husk produced in the factory, in turn dictated by the overall consumer demand for soluble coffee, and also on the factory's overall energy requirements. Given current production volumes and market demand for the goods produced, it is expected that in the coming years the biomass boiler will generate between 15 and 20 per cent of

Table 4.2 Energy produced from coffee husks, 2010–2012

Year	Coffee grounds (tonnes)	Husk-generated energy (GJ)	Husk-produced energy used in the factory	Factory total energy requirements (GJ)
2010	4,537.4	35,615.4	31%	114,888.9
2011	3,105.69	24,379.7	12%	203,164.1
2012	4,351.64	34,160.4	15%	261,069.3

Source: Nestlé records

the factory's energy requirements (Nestlé Colombia, 2013b). This project has contributed to significant energy savings for the factory, ensured cleaner production processes and reduced solid wastes that need to be disposed of in environmentally friendly ways.

During the last five years, energy consumption has fallen by 29 per cent and greenhouse gas generation has declined by 47 per cent. It should be noted that despite higher factory energy consumption, water and energy requirements per tonne of produced goods has steadily decreased during the last seven years. In 2007, 13.78 GJ were needed per tonne of final output. By 2012, this number had fallen to 9.5 GJ/t. This represents a 31 per cent reduction in energy consumption (Nestlé Colombia, 2013b).

Coffee waste has dramatically fallen as it has been turned into clean energy. Yet, there is still a considerable volume of residues left over from the operation of the biomass boiler. Tests have been successfully performed to turn the residual ashes into organic fertilisers that coffee growers could potentially use. Unfortunately, as of now, qualified suppliers have not been found to bring this project to scale. The initiative has been temporarily put on hold even when it is worth re-developing given the high costs of fertilisers in Colombia. Additionally, adequate, opportune and necessary fertilisation plays a defining role in obtaining high crop yields. The production of coffee residue-based fertiliser remains a potentially profitable and beneficial innovation to enhance productivity in coffee plots.

Other innovations have also contributed to the reduction of the plant's environmental footprint. Take water for instance. According to Colombian legislation, brown-coloured wastewater resulting from the coffee washing process can be discharged directly into water streams. In 2006, Nestlé Colombia made an additional investment of USD 3 million to improve its water treatment plant and remove the harmless, yet unpleasant, brown colour that results from coffee processing. Additionally, during the last five years, the company's water footprint has substantially improved. Consumption has decreased by 46 per cent for each tonne of product manufactured and wastewater generation has declined by 17 per cent per tonne.

The company's corporate culture requires that water must be used efficiently. Beyond that, there is top most pressure for production processes to be more water efficient. Progress is tracked through an indicator signalling water use per tonne of product and water extracted per tonne of product. Not only should factories reduce their water requirements, they have to continually increase the efficiency of their production processes.

As the company improves its global technical knowledge about water footprinting and water management, Nestlé passes on that expertise to local factories and their surrounding communities. There are specific on-site factory actions to build a more judicious water culture. For instance, as part of the SuizAgua initiative, the entire life cycle for dairy production was analysed. It first identified the areas of opportunity, after which changes were introduced to water management systems and production lines. The result has been more efficient and less water intensive manufacturing processes.

Employees are also expected to modify water consumption habits in and outside the factory. In an attempt to change community-wide water management systems, to 2012, 160 girls and boys had taken part in educational activities surrounding the importance of having access to clean water. Additionally, water management capacities have been strengthened for some 100 producers. These activities contribute to societal behavioural changes that may lead to favourable long-term aggregate impacts on communities.

Stability anchors

The private sector has the potential to create wealth and promote socio-economic development by generating employment and contributing to financing public goods and services through their fiscal contributions. However, whether we refer to micro and small-scale firms or to large transnational corporations, conflict and structural violence acutely lessen the potential benefits host communities could otherwise reap. According to the Conflict Index in Colombia (ICOC), between 1964 and 2011, the country's annual growth was on average 0.59 per cent lower and per capita GDP 0.45 per cent below what it would have been in a more stable and peaceful situation. These losses were magnified in 2000, when violence intensified. That year, GDP was 1.9 per cent below its potential rate of growth and per capita income 1.4 per cent lower (Riveros Saveedra, 2013).

Nestlé cannot insulate its operations from conflict in the immediate vicinity of its facilities, and the violence affecting its staff, and its suppliers. In 2007, a bomb was placed in a shortly-to-be-inaugurated milk pre-condensing factory in El Doncello, and also in the plant in La Florencia, both of them located in the region of Caquetá. Over 400 soldiers had to guard the company facilities. This made it a particularly difficult time for Nestlé in the seven decades that it has been operating in Colombia. The attack to the El Doncello factory hampered the collection of over 70,000 litres of milk, directly affecting 1,500 small milk producers (*El Tiempo*, 2007a; Ornelas, 2007). In Florencia, the company plays an important economic role given the region's economic focus on milk and beef production, as well as the cultivation of bananas, yucca and coffee for export. In Florencia alone, Nestlé buys 300,000 litres of milk every day (Las Páginas Amarillas de Florencia, 2014).

Following these attacks, Nestlé has sought ways to prevent future incidents, protect its workers and maintain a functional productive chain. Despite disruptions to its local supply chains and security concerns for the factory workers, the company decided to engage more strongly than ever with the local communities. It is determined to remain in the conflict-affected areas and continue to act as a regional development engine and a source of employment. This unwavering and robust corporate presence has in fact accelerated the delivery of social investments that can improve the region's social investment climate (Nestlé Bugalagrande factory staff, personal communication, 6 June 2013).

These commitments have been globally acquired and locally honoured. Nestlé is a founding member of the United Nations Global Compact, the largest voluntary corporate responsibility initiative in the world. It has agreed to uphold ten principles covering human and labour rights, environment and governance. Presently, the initiative encompasses 8,000 companies in 140 countries and over 4,000 civil society organisations collaborating with firms in tackling common issues or acting as accountability watchdogs.

The Ten Principles of the United Nations Global Compact list a series of ten commitments businesses are to embrace, support and implement on issues regarding human rights, labour, environment and governance. Companies are to: (1) support and respect the protection of human rights; (2) make sure they are not complicit in human rights abuses; (3) uphold freedom of association and collective bargaining; (4) eliminate forced labour; (5) abolish child labour; (6) eliminate discrimination of employment and occupation; (7) support a precautionary approach to environmental challenges; (8) promote environmental responsibility; (9) encourage the development and dissemination of green technologies; and (10) work against corruption (UN Global Compact, 2014).

This voluntary regime seeks to further sustainable business models and markets and to deliver long-term benefits to the communities connected to its activities (UN Global Compact, 2013). Committed companies are expected to make the UN Global Compact principles part of their business strategy, decision-making processes, reporting criteria and basis for partnership building (UN Global Compact, 2014). These broad development objectives are to be upheld under all circumstances, especially at times when violence breaks down institutions, social capital, stability and the prospects for prosperity. Unfortunately, detailed studies assessing the community impact of private sector investment in productive sectors during complex emergencies are neither available for Colombia nor Nestlé's activities in the country (Nelson, 2000). Systematic studies gauging the impacts of private sector enterprises on maintaining stability at times of conflict constitute another gap in academic literature. This is a research area of rising relevance for countries, academics, universities, development institutions, think tanks and civil organisations interested in conflict resolution and peace building processes.

This research team gathered qualitative data and individual and collective anecdotal evidence revealing the important role Nestlé's factories have played in some of the conflict-affected areas in Colombia. It also collected information concerning the important challenges the company faced to manage the physical security of its assets and staff members. The findings reported in this document focus on coffee growers selling their produce to Nestlé and the coffee-processing factory in Bugalagrande. Yet, the complexity of Colombian socio-political geographies, and the conflict itself, suggests the company has had to take on a flexible approach to safety, security and business. Adaptability is a crucial trait when operating simultaneously in different regions of the country, each one with different conditions and requirements.

Academics have tried to find the causal link between violence and poverty and whether opportunism (greed) or injustice and exclusion (grievance) ignite conflict (Collier and Hoeffler, 1998, 2004). However, whether poverty is a conflict trigger or its consequence, the fact is that the poor are disproportionately affected by it (Duranton and Sánchez, 2005). In Colombia, the intensification and expansion of conflict further weakened the justice system, respect for the rule of law and the protection of already feeble property rights. It also interrupted the adequate functioning of the local and national institutions, electoral channels to voice discontent, political parties and the allocation of tax revenue to fund public services. Violence also posed additional difficulties for the delivery of public goods and economic activities, compromising employment opportunities and growth (Galindo et al., 2009).

Beyond the individual and collective tragedy that has resulted from the loss of life, security, certainty, assets, social fibre and safety, the protracted conflict in Colombia has also affected the inter-temporal economic decisions of society as a whole. A vast body of literature exists on the role multinational enterprises involved in extractive industries can have and are collectively looking to play in vulnerable societies. Such firms provide large and badly needed revenues to governments. However, much concern has been expressed regarding the use of taxes and royalties made to host countries with feeble governance systems or to non-state actors given the potential contribution they can make to finance violence and increase the conflict's economic stakes. Investment in complex emergencies is thus 'one of the most sensitive international investment issues of the early 21st century' (OECD, 2002: 5). For more discussion on these issues see: Sullivan (2003); Collier and Hoeffler (2004); Hillemanns (2003); Bennett (2002); OECD (2002).

In the most severely affected areas, violence has shaped people's decisions and behaviours regarding investment, consumption and savings. It has also affected social, political, economic, legal and cultural institutions and all types of social infrastructure across the country (Leiteritz et al., 2009). As it is to be expected, the resulting changes in the economic landscape have lasting repercussions in the current and future potential for growth. The contraction of the legal productive sector, the uncertain and inefficient supply of inputs and factors of production, as well as the depreciation of and disinvestment in physical and human capital delays long-term post-conflict economic, social and institutional recovery (Galindo et al., 2009). Attempts at maintaining normality and normalising social, inter-relational, commercial, labour and economic activities thus help to mitigate the otherwise deterioration of norms, institutions and quality of life (Collier and Hoeffler, 1998, 2004; Rettberg, 2010). One such effort is exemplified by the long-standing presence of multinational enterprises, like Nestlé, in troubled countries, like Colombia.

Moreover, partnerships with the local government and non-governmental organisations, international development agencies and civil society groups have strengthened the locale's social capital. This type of joint work has led to the development of innovative strategies that has helped alleviate problems in conflict zones by improving the inhabitants' earning capacity. From 1986 to 2002, Nestlé Bugalagrande worked in partnership with the Sarmiento Palau Foundation to support entrepreneurial activities, social development initiatives and poverty alleviation efforts (Fundación Sarmiento Palau, 2009). Regretfully, the actual outputs, outcomes, results and impacts of such activities have not been adequately documented. The lack of knowledge about the effectiveness of such initiatives limits its potential to be scaled-up and replicated, if considered successful.

Nestlé has leveraged its economic impacts in the surrounding municipality of Bugalagrande to promote social stability. By continuously contributing to its workers and suppliers' standards of living, the Bugalagrande factory has widely shared the benefits of the company's commercial success, reducing the prospects for violent conflict and dispersing tensions (Carnegie Commission, 1997; Davy, 2001). According to testimonies gathered during a field visit to the factory in May 2013, the company's regular activities during the years of heightened violence served to normalise life in the municipality. Bugalagrande's Mayor, the Director of the Chamber of Commerce of Tuluá, local entrepreneurs and people from the area all agreed that the company brought about security and a relative tranquillity to an area where 'violence was becoming commonplace to the point of being seen as part of the landscape' (Durán Castro, Chairman of the Board of the Foundation for the Integral Development of Valle, personal communication, May 2013).

Beyond economics, companies ought to be sensible to local realities and the role employees play in their communities. The culture of violence and collective grievances in many of the regions where the company operates urged and encouraged it to get closely involved in local peace-building processes, which ultimately would have positive influences on the company's performance. In January 2007, The Revolutionary Armed Forces of Colombia (FARC) attacked the company's milk trucks and collection centres in Caguán. In the Doncello and Florencia factories two car bombs exploded and the FARC stated that anyone selling milk to Nestlé was to be considered as a military target (*El Tiempo*, 2007b, 2011).

The situation deteriorated to the extent that operations in the affected areas had to be suspended. This noticeably jeopardised the community's livelihoods, as milk producers could not sell their product to the company and there were no other economic alternatives that could offer a similar level of income. Yet, to the surprise of many in Nestlé, the population did little to denounce the attacks (Fundación Ideas para la Paz, 2011). That same year, the company and Sinaltrainal union were becoming antagonistic players in Bugalagrande.

Corporate relations with the surrounding community turned tense despite Nestlé being a local economic engine and the main employer for people in the town. Strained relations with workers quickly reflected onto the Bugalagrande community at large. Deteriorating dealings with Sinaltrainal also created a dent between Nestlé and its employees. This prompted the company to re-conceive its ties with its workers and suppliers and the local populations. The management at the Nestlé Colombia factories took the decision to become part of local efforts to foster peaceful conflict resolution.

In retrospect, this corporate approach made good business sense. On one hand, the company was setting the foundations to restart operations in previously violence-affected areas in Caquetá and building a space to dialogue and improve relations with its workers' unions. On the other hand, the initiative was to help the company to do well by doing good, despite adverse circumstances. As the company deepened its understanding of the environment where it is located, it built the required social legitimacy and desirability to operate in those communities (Fundación Ideas para la Paz, 2011). For instance, it supported the creation of peace fostering centres.

The company has sponsored and developed an initiative to establish Peace and Reconciliation Centres in Bugalagrande (Valle del Cauca), San Vicente del Caguán and Cartagena del Chairá (Caquetá). Such schemes target victims of the conflict, populations vulnerable to being forcibly displaced, youth at the risk of conscription by illegal armed groups, socio-economically challenged groups and the population in general (Emprender Paz, n.d.). In Bugalagrande, Nestlé has partnered with the Foundation for Reconciliation to establish a Reconciliation Centre. The aim is to tackle community violence and build social capital.

With an initial investment of COP 80 million, the Centre was inaugurated on 28 September 2010 as a meeting point for all population groups to come together and foster peaceful social interactions. This public space is seeking to rebuild community trust by teaching individual and collective mechanisms to deal with conflicts, disagreements and grievances without resorting to physical or verbal violence. It has also had important, unexpected and positive repercussions in improving household dynamics and addressing domestic violence (Emprender Paz, n.d.; Nestlé Colombia, 2013b).

Both in Bugalagrande city and in Caquetá, where the second Centre has been established, violence has become almost structural and at times even validated by the community itself (Emprender Paz, n.d.). Initially, this normalised aggression posed significant attitudinal challenges to the process leading to the opening and functioning of the Reconciliation Centres in those two municipalities. With time, the Centre and the people who work and volunteer there have sparked behavioural and discourse changes. The most encouraging developments have been identified amongst vulnerable groups in the community. Socially, the project proved so successful that it was included as part of Bugalagrande's public programme to build a culture of peace. This entrenchment into public policy measures assures the sustainability and continuity of the initiative. It also illustrates how possibilities to create shared value are intrinsically dependent on a country's prevailing political landscape, economic policy and social reality.

For Nestlé, the Centre opened up new communication channels with the community, allowing both the local populations and the corporation to see each other as strategic and interdependent actors (Fundación Ideas para la Paz, 2011). In 2011, the initiative received an award by the '*Emprender Paz: La apuesta empresarial*' organisation for its contributions to building peace in the country. The award is trying to bring attention to the different initiatives the entrepreneurial sector in Colombia undertakes to address and mitigate the causes and effects of conflict in vulnerable and affected populations (Emprender Paz, n.d.).

The *Emprender Paz* organisation was formed to highlight the role of the private sector in creating scenarios for sustainable peace. In 2008, the group listed a number of companies engaged in peace building projects or in supporting activities protecting affected and vulnerable populations. The organisation is currently supported by the German Society for International Cooperation (Deutsche Gesselschaft für Intenationale Zusammenarbeit, GIZ), the Konrad Adenaeur, Social and Nogal Foundations, the Swiss Embassy to Colombia, the National Learning Service (SENA), *El Tiempo* newspaper, the Social Responsibility and Sustainability International Centre (RS) Magazine, Colombia's Newsagency Colprensa, Aviatur Airlines and the United Nations Development Programme (UNDP).

Nestlé cannot simply ignore the events, people, circumstances and public policies upon which its long-term success depends. It certainly cannot overlook union strife in its coffee processing factory in Bugalagrande, the country's overall political situation, guerrilla attacks to its facilities and employees, declining output levels for key crops, environmental deterioration of the areas neighbouring its factories and low levels of human capital and capacity development. Doing so would decrease factory output, cause employee dissatisfaction and resignations, diminished production capacity, shortages of key inputs, failures to attract skilled workers, etc.

Peace-building programmes offer another example of how Nestlé, as a private company, is prospering along societies, and not at their expense. Reaching the balance between corporate aspirations and societal expectations is an ever-changing act. Companies have to continuously fine-tune their approaches in order to meet dynamic demands put forward by multiple stakeholders.

Agents of change

Seen as a positive 'agent of change' and an 'engine for rural development', the Bugalagrande factory has been consistently acknowledged for the advantageous externalities, beneficial quality and human capital spillovers, as well as favourable multiplier effects it has brought about in the region. Reflected in Nestlé's records and the collective memory of the municipality's inhabitants and factory workers, from the time when the Bugalagrande factory opened, the economic and social profile of the town has changed remarkably. A mainly agricultural township, during the 1970s, poverty rates were as high as 58 per cent and malnutrition affected around 10 per cent of the population. Since then, poverty incidence has fallen by 20 percentage points and malnutrition rates have been significantly brought down to only 3 per cent (Nestlé Colombia, 2013b). These are improvements to which the Swiss company has contributed via employment generation, the production of fortified foods and nutrition-oriented campaigns.

For over five years now, the Nestlé Healthy Kids programme has been put in place in the geographical areas where the company is present. The aim is to encourage and further strengthen healthy living habits amongst children and have them take that knowledge and new behaviour to their homes to improve household habits. This scheme is implemented in the company's sponsored toy libraries, or *ludotecas*, where beneficiary kids are offered a safe space for social development. There, and in partnership with the Childhood Day Corporation, 15,000 children from the region learn about healthy living, receive basic nutritional education and become responsible water consumers.

The Toy Library initiative implemented in Bugalagrande (Valle) was inaugurated on 19 September 2007 (Government of Valle de Del Cauca, 2007). On 6 February 2013 the company opened a second facility in Dosquebradas (Risaralda), as a shared project with the Municipal Government and the Childhood Day Corporation. A third one is operating in Florencia, to the benefit of some 600 children and their families. In the last five years, 85,000 people (children and their families) have taken part in the activities in the Bugalagrande and Florencia *ludotecas* (Nestlé Colombia, 2013b). Equally focused on the young generations, the Adventurers Club, a scouting association, teaches children about civic, civil, social, convivial, environmental and sporting values. Two of its former members have become Swimming Pan-American Champions (*El Tiempo*, 2007c).

Beyond these factory-specific social initiatives, Nestlé supports a series of community projects all over Colombia, mostly in support of the company's own areas of expertise and comparative advantage: wellness and nutrition. It has implemented, supported and financed programmes for the elaboration of fortified cookies, the operation of a Nestlé travelling vehicle offering nutrition-related workshops and counselling (*Nutrimovil*), the creation of household gardens and support for welfare homes, etc. All these initiatives aim at bringing about positive changes in the lives of the people, starting from good nutrition (Industria Alimenticia, 2009). To attain these goals, Nestlé has created alliances with a wide range of public, private, not-for-profit and non-governmental actors.

On the nutritional front, the company has supported the '*Nutrir*' Foundation since 1991. In 1981, a group of businessmen established the foundation in an effort to tackle malnutrition in Bogotá. In 2012 alone, this scheme supported 285 children, pregnant mothers, and parents (Genesis Foundation, n.d.). It has been estimated that in the last 15 years of operation, this programme helped 2,000 kids, below the age of 12, to overcome acute malnutrition in Colombia's capital city (Portafolio, 2007).

Nestlé also offers in-kind and financial support to similar initiatives. Since 2001, the company has been making in-kind yearly donations for USD 35,000 to the Archdiocesan Food Bank, which services around 70,000 people (Abaco, 2013). In 2004, children assisted by the BareFoot Foundation in the neighbourhood of Altos de Cazuca, Bogotá, began receiving nutritional support (BareFoot Foundation, 2008). In addition, from 2006 and for two years, the company contributed to the Bugalagrande's Children's Breakfast Programme. The scheme was eventually discontinued by the department's administration in 2008. Similarly, Nestlé Colombia has partnered with the National Institute for Family Welfare to produce and supply fortified crackers for over 500,000 children at high risk of malnutrition in 600 localities. Moreover, the multinational is supporting the Food Security, Food and Nutrition Improvement Plan in Antioquia (MANA by its acronym in Spanish) aimed at improving the access the vulnerable populations have to sufficient and nutritious food (Government of Antioquia, n.d.).

Thanks to this type of strategy aimed at addressing under- and malnutrition, developing affordable and accessible fortified foods for low-income consumers and improving nutritional quality, Nestlé has ranked as one of the top three performers in the Access to Nutrition Index (ATNI). This index ranks the 25 world's largest food and beverage manufacturers according to their policies in governance, products, accessibility, marketing, lifestyles, labelling and engagement (Nestlé SA, 2013a; Global Alliance for Improved Nutrition, 2013). In a similar fashion, in 2013 Nestlé was the top performer in Oxfam's Behind the Brands sustainability scorecard, obtaining 45 out of 70 possible points. The organisation ranks among the top ten food and beverage companies with strategies and actions in place to improve food security and sustainability. Their performance is evaluated in areas such as transparency, farmers, women, agricultural workers, access to land, water and climate change (Oxfam International, 2014).

In Dosquebradas, Bugalagrande and Florencia, Nestlé has partnered with Maloka, the country's first interactive science and technology centre, to develop a travelling fair and computer game, the Nestlé adventure, promoting healthy eating habits (Colombia Tecnología, 2009). The firm has also collaborated with the United Nations Children's Fund (UNICEF) and the Secretary of Education in Bogotá to provide school supplies (Edukits) to children from impoverished backgrounds and on the verge of dropping out of school (Nestlé de Colombia records). For seven years now, since 2007, Nestlé has made contributions and donations to the Juan Felipe Gómez Escobar Foundation, an institution caring for poor children and adolescent mothers and their children in hospitals around the region of Cartagena (Fundación Felipe

Gómez Escobar, 2014). From 2006, it has also worked with the Swiss Foundation 'Help for Children', which provides shelter for needy and orphaned children between the ages of one to six.

The multinational has also joined efforts to preserve Colombia's cultural heritage. It is one of the Corporate Friends of the Bank of the Republic Art Collection. Moreover, it helped the Productive Hands Cooperative – a recycling co-operative for unemployed women – to acquire plastic processing machines. This organisation now handles cardboard and other waste materials produced in the Dosquebradas factory, with direct and indirect benefits for 75 families (Nestlé de Colombia records; *El Tiempo*, 2008). All of these efforts strengthen the company's own position in the country. At the same time they help to improve the quality of life of vulnerable communities (Portafolio, 2007).

Sustainability is a global effort

As a global company, Nestlé is so large in profits and jobs generated, countries where it is present, number of factories, range of goods produced that it needs to have a good understanding of the numerous factors that impact on its activities. Although sustainability is only one of them, it is crucial to keep the company's actions viable and profitable in the long term. As such, Nestlé has integrated sustainability considerations, indicators, strategies and action plans into its core operations at global and local levels. The articulation of global goals and local actions to attain them reflects on the company's overall performance as reflected in a number of leading environmental and sustainability rankings and indices.

For example, the company has recently gained recognition from the Dow Jones Sustainability Index (DJSI), which is the world's first and leading global sustainability benchmark. S&P Dow Jones Indices LLC and RobecoSAM AG jointly developed the Dow Jones Sustainability Indices. RobecoSAM was founded in 1995 as a Sustainability Investing specialist and has since then gathered one of the most inclusive sustainability databases in the world. Information is gathered on an annual basis and environmental, social and corporate governance (ESG) investment assessments carried out for over 2,500 publicly listed companies (RobecoSAM AG, 2013). The DJSI acts as a guiding tool for investors interested in integrating economic, social and environmental performance criteria into their portfolios. For companies, it provides an engagement platform for companies to adopt sustainable practices (Dow Jones Sustainability Indices, 2013). This weighted average awards 40 per cent importance to economic dimensions, 29 per cent to the environment and 31 per cent to social issues.

In September 2013, Nestlé positioned itself as the 2013–2014 Industry Group Leader in the Food, Beverages & Tobacco category. With an 88 per cent score, the company's performance was double of the industry average. Environmentally, it obtained 97 aggregate points. This was the best overall score amongst all evaluated companies and also the highest performance in three out of four parameters gauged: mitigation of water related risks, environmental reporting and operation eco-efficiency. In the fourth category, raw material outsourcing, Danone outdid Nestlé despite the company scoring twice as high as the DJSI Industry average (RobecoSAM AG, 2013).

Nestlé outperformed the DJSI Industry average in 2013 and 2014. This suggests the company has been doing well in terms of sustainability practices vis-à-vis other companies. For instance, Nestlé's strategy for emerging markets was the best in the industry. However, it can still catch up with the best companies in the Food Products Industry in the DJSI economic and social dimensions. These parameters allow companies to benchmark and compare their performance in relation to that of their competitors. It also signals the areas to which companies may have to pay closer attention. In 2014, Nestlé was slightly outdone by Danone in supply chain management and surpassed by a wider margin by Unilever in terms of corporate governance in health and nutrition. Socially, Nestlé may have to intensify efforts to attract and retain talent, develop human capital development and improve on occupational health and safety. It was in these three key indicators in which it was outdone by the widest margins.

Also on the environmental front, the company obtained the maximum score (100 points) in the Carbon Disclosure Project (CDP) Climate Disclosure Leadership Index and Climate Performance Leadership index for the second year in a row. Published in the Global 500 Climate Change Report, both Indices are used by investors as a proxy capturing corporate climate change management practices. They compute the efforts to curb carbon emissions and increase transparency of the disclosed information of the top 500 companies in the FTSE Global Equity Index (Carbon Disclosure Project, 2014). Nestlé is also included in the FTSE4Good Index Series, a measure launched in 2011 to assess how companies perform in terms of corporate responsibility standards in areas related to environmental sustainability, positive relationships with stakeholders, upholding and supporting universal human rights, ensuring good supply chain labour standards and countering bribery (FTSE International Limited, 2006; Nestlé SA, 2014b).

Regarding water, the company has participated on the CDP Water Programme every year since the initiative was launched in 2010. Nestlé thus supports the project's efforts to promote sustainable corporate water stewardship and address global water challenges. As part of the Water Programme, the company works to achieve water efficiency across operations; improve suppliers' water management practices and save water in the company's upstream supply chain. Acquired commitments also include raising awareness of water access and conservation amongst employees, communities and consumers. Participating companies are expected to report on attained progress (Carbon Disclosure Project Water, 2013; Nestlé SA, 2013b).

Since 2003, Nestlé factories all over the world have improved their energy efficiency, switched to cleaner fuels and invested in renewable sources. The scale of these efforts is considerable. The company has 447 factories in 86 countries and in most of them greenhouse gas emissions per tonne of finished product have been halved or substantially reduced. This type of global environmental stewardship reflects upon and is only possible due to factory level performance. Colombia, the country on which this study focuses, has contributed to the DJSI score through the SuizAgua Colombia project, the Cisco plant, the Nescafé Plan and the Milk Development Plan in Caquetá.

SuizAgua is a Swiss Agency for Development and Cooperation (SDC)-led initiative to measure the water footprint of Swiss companies in Colombia. The public–private project has brought together Nestlé, Clariant, Holcim, Syngenta and seven Colombian companies to test the draft international standards for water footprinting: ISO 14046 Water Footprint – Principles, requirements and guidelines. The norm was used to analyse each life cycle stage, from supplying dairy farms to the final product leaving the factories. The information was used to estimate the water footprint for one tonne of dairy products manufactured in the Florencia and Bugalagrande factories.

This exercise pointed to the most water intensive steps in dairy production, namely energy consumption and the milk supply chain. In response, Nestlé Colombia took actions to protect water springs and manage waste on 95 farms. At its plants, it implemented leak control measures, installed steam recycling systems and standardised practices for production stoppage and cleaning periods. In only four years, from 2009 to 2013, water extracted per tonne of dairy product was reduced by 44 per cent in the Florencia plant, from 4.9 to 2.7 cubic metres per tonne. During that same period, electricity consumption decreased from 96 to 62 kWh per tonne of product, which represents a decrease of 35 per cent (Nestlé SA, 2014b).

Moreover, the Integrated Water Management Programme Nestlé initiated in procuring dairy farms in Montañita and Morelia, Caquetá, has led to a 7 per cent decrease in the company's water footprint. In terms of improved supply chains, the sustained implementation of the Milk Development Plan in Caquetá has supported 1,936 dairy farmers who now directly provide Nestlé with more than 50 million litres of milk every year (Nestlé SA, 2012a). Productivity has increased by 400 cubic centilitres of milk per animal every day (Portafolio, 2007). The two coffee-related initiatives, the Cisco Project for the reuse of coffee residue and the Nescafé Plan for sustainable sourcing are discussed in detail later.

Clearly, no one company can excel in absolutely all sustainability parameters, especially since performance is context determined and sustainability is a moving goal. Yearly changes in rankings across different criteria evaluated by indices such as the DJSI, CPD or FTSE4Good are just a small reflection of this. They also show that companies do engage in a race to the top as they seek to build strong corporation-wide capacity to implement, scale up and sustain their own initiatives. Multinationals have promptly understood that in a time when sustainability is defining competitiveness, 'companies that take the lead on sustainability will be market makers rather than market takers' (World Economic Forum, 2009).

Big companies, Nestlé included, have additional vested interests to publicly and third-part verifiably demonstrate significant progress in raising sustainability performance levels and closing planning–implementation gaps. Potential and actual shareholders, traditionally the most powerful group in a company, reward such engagement. Investment communities are gradually factoring in environmental stewardship, social responsibility and good governance in their calculations gauging a company's long-term value (Guthrie, 2014). They have reacted positively to Nestlé's local and global targeted policies and actions aimed at improving corporate sustainability. The company's market capitalisation has nearly reached CHF 204 billion and sales in 2012 amounted to CHF 89,931 million (Bloomberg, 2013).

However, a race to the top does not exclude co-operation between the most important players in the global food industry. The mission and vision statements of many companies in this sector clearly spell out their commitments to comprehensively 'promote sustainable agriculture' and make of it a mainstream practice. To do so, they join efforts. Companies know that, regardless of market share and size, no single one of them can normalise the mechanisms, practices and principles that can simultaneously account for the economic, environmental and social aspects of agricultural activities all over the world. Instead, they work together. The food industry has created inclusive platforms for large and influential private players to consider the long-term supply of agricultural inputs, potential degradation of natural resources this production may result, the asymmetries of global supply chains, and the consequences these imbalances may have for the development and wellbeing of rural communities (Nestlé SA, 2008).

In 2002, Nestlé, Danone and Unilever founded the Sustainable Agriculture Initiative (SAI). This platform was set to: (1) address quality and safety problems in the food supply chain that may affect consumer confidence in everyday products; (2) manage the changes in food demand that will emerge from increasingly affluent societies, population growth in emerging economies and changing diets and lifestyles; and (3) foster agricultural productivity by preventing and mitigating adverse impacts their activities may have on the environment and on the use and management of natural resources (SAI, 2013).

According to the SAI Platform manager Didier Lebret, Nestlé's experience in coffee has sped up the development of Coffee Principles and Practices within the SAI working group on this crop (Nestlé SA, 2006). It has also allowed SAI to learn from the results of pilot projects testing the SAI Platform practices in Latin America (Nestlé SA, 2006). These efforts have important potential for scalability and the engagement with an extensive range of stakeholders. Intra- and inter-firm knowledge exchange, partner inclusiveness and the potential to sizeably replicate successful experiences in different countries are proving determinant factors in spreading and engraining sustainability practices in the food industry.

Nestlé's own internal activities show the extent to which intra-firm activities are increasingly adopting SAI principles. In 2007, sustainability-oriented programmes covered 28 of Nestlé's markets. By 2013, SAI programmes framed the commercial activities in 46 of the company's markets. The implementation of responsible sourcing programmes means that 100 per cent of Nestlé's direct procurement markets followed sustainability principles (Nestlé SA, 2008, 2014d).

★★★

To feed its production processes, similar to all other multinationals, Nestlé procures vast amounts of raw materials, unfinished and semi-unfinished products and services. Most of these inputs come from developing countries. This is a position companies can leverage to foster sustainable economic development in supplying markets, strengthen local communities and contribute to alleviate poverty. At the same time, and in response to external pressure and internal strategic management, companies can also use the global scope of its operations, market power and international expertise to stimulate value added production (Doh, 2006).

Since the company set up a factory in Colombia in 1944, its management and people in charge have tried to make it be as socially and environmentally responsible as it has been economically successful. That approach has prompted changes in processes leading to more sustainable agricultural practices that, in turn, have generated social, economic and environmental benefits. Directly and indirectly, those regions in the country linked to Nestlé are now engaged in higher value-added production. Farmers have higher incomes, producer associations have flourished and agricultural opportunities in a mostly unprofitable sector have been created. In addition, the company's internal codes of conduct and sustainability commitments have contributed to the conservation and better management of scarce natural resources.

With rapidly changing national and global scenarios, this is clearly an evolving process in which the company has to continuously look for ways to have a more dynamic, responsive and regular dialogue with the communities where it is based. Better enterprise–society communication channels can help plan, implement and evaluate social and economic programmes so that they have a wider reach and impacts, larger scale and take a more systematic approach to creating shared value. Besides, it is an important way of shaping and mobilising social opinions and concerns, both locally and internationally (Chairman of the Chamber of Commerce of Tuluá, personal communication, 2013).

Nestlé's value chain in Colombia rests on two inputs and final products that merit particular attention: milk and coffee. The establishment of dairy districts is one of its areas of expertise par excellence whilst coffee ties global consumers to local producers in one of the countries with the strongest coffee cultures in the world. Every year, the company buys some 151 million litres of milk from 2,600 small dairy farmers, positioning it as the fourth largest milk buyer in Colombia.

Colombian coffee beans, on the other hand, are key for the world's largest coffee purveyor and coffee brand (Nestlé Colombia, 2013b). Worldwide, the company buys nearly 14 million bags of coffee, 60 kg each, out of which 1.2

million bags come from Colombia. These purchases of Colombian Arabica Milds represent 15 per cent of the country's total coffee production. This makes Nestlé the largest private coffee buyer in the country. Around 85 per cent of the coffee bought is exported to other Nestlé factories around the world, mostly for consumption in the United Kingdom, Russia, Japan and Germany (Andacol, 2013). It is thus not surprising that coffee and Nestlé are so closely interlinked. The crop holds a crucial place in Nestlé's growth strategy in Colombia, in turn Nestlé's actions in the coffee trade are echoed in the local chain as a whole, making it an impact enterprise in the country's coffee sector.

The company is mobilising its corporate strength and core capabilities to ensure its suppliers adhere to the same codes of conduct that the company endorses, champions and strives to put into practice. As part of these efforts to source responsibly, and create shared value for the company and the different tiers of shareholders alike, the firm is helping coffee growers in Colombia to increase productivity, use the resources available to them more efficiently and safely, diversify their economic activities and improve the quality of the produced raw materials. These activities have been awarded corporate priority as part of global commitments but are being implemented and managed locally. Subsequent sections of this book will carefully assess how this process has unfolded for Colombian farmers choosing to sell coffee beans to Nestlé.

Note

1   1 USD = 2,017.82 COP to 17 February 2014 (Bloomberg, 2014).

5   From farm to cup

The Nescafé Plan

Every second, some 5,500 cups of Nescafé® are drunk in the world. More than half of this soluble coffee is processed in developing countries. Iconised by a red coffee cup, the brand targets ambition flirters, success seekers, young cosmopolitans and people on the move. Nescafé® is also the world's biggest selling coffee brand by a factor of five, holding the largest global coffee footprint with a presence in 180 countries (Nestlé SA, 2014d). On the premium side of the coffee spectrum, Nespresso, an autonomous globally managed business of the Nestlé Group, created and revolutionised the premium portioned coffee market. Founded in 1986, the brand's direct-to-consumer business model and route to market has shaped the global coffee culture. It is thus not surprising that coffee holds a crucial place in Nestlé's strategy for rural development and the improvement of the living conditions of those supplying this main input. Framed within its 'Creating Shared Value' business strategy, the company has put in place a coffee-specific agenda to streamline and optimise the company's coffee supply chain. Its two components are the Nescafé Plan and the Nespresso AAA Sustainable Quality™ Program.

Launched in August 2010, the Nescafé Plan has emerged as a holistic, long-term global initiative that covers several countries, one of them Colombia. Its objective is to boost the competitiveness and attractiveness of the local coffee sectors in procuring markets. In its initial and first phase, which is to continue till 2020, the initiative has made the public commitment to invest a total of CHF 500 million in coffee projects all around the world. For a decade, CHF 350 million will support Nescafé® and CHF 150 million will be used to strengthen the Nespresso Ecolaboration™ platform. These schemes are building 'market-based incentive systems to improve the environmental and social impacts of coffee farming, processing and trading' (Perez-Aleman and Sandilands, 2008). They are a way of securing a market for coffee supply grown using sustainability practices.

Activities comprised in the Nescafé Plan are grounded on the premise that sustainable practices lead to higher productivity and output, which generates higher incomes (Nestlé Nespresso SA, 2013b). The initiative is working to change farming practices and lives by identifying and then promoting the drivers of sustainable production and processing. It is also based on the understanding that increasing and maintaining productivity, boosting incomes and improving living conditions are incremental processes, and the result of sustained and constant efforts.

The Plan is centred on developing long-term relationships with the coffee farmers that sell their coffee beans to the brand. This is to be achieved by strengthening the capacities in agriculture and other relevant and associated areas. One notable way in which relationships between Nestlé and its community of farmers in Colombia are built and strengthened through extensive and intensive technical assistance programmes and direct procurement operations is called Farmer Connect. Every year, farmers all around the world benefit from advice on farming and post-harvest practices and processes. The previous chapter describes Colombia's unique coffee institutionalisation and the technical assistance services coffee producers have traditionally received. The company has built on this existing scheme and has pushed to intensify and improve the agricultural support services it provides. It has combined technical assistance with connectivity to global supply and value chains. Further sections describe the technical assistance component of the Nescafé Plan.

Globally, the aim is for Nescafé® to make its green coffee purchases directly from farmers and their associations; this means buying 3,180,000 tonnes from 170,000 farmers, between 2010 and 2015. It also comprises the rejuvenation of old coffee plantations with 220 million high-yield, disease-resistant plantlets by 2020. The Plan is also working towards deepening Nestlé's work with the Rainforest Alliance, the Sustainable Agriculture Network (SAN) and the Common Code for the Coffee Community (4C Association) to eradicate environmentally harmful and socially unacceptable practices and to improve farming activities.

Globally, the number of coffee farmers receiving technical assistance and training through the Nescafé Plan has more than doubled since 2011. That year, 56,994 coffee growers benefited from this service. In 2013, close to 124,570 farmers received technical support. This represents a considerable increase of 22.58 per cent from the 101,622 growers who had access to this service in 2012 (Nestlé SA, 2014b). In 2013, a team of 200 agronomists took to the task of visiting 30,039 farms to deliver technical assistance, distribute plantlets and build local capacities.

In Colombia alone, the first stage of the Nescafé Plan was rolled out in four municipalities in Valle del Cauca – Andalucía, Bugalagrande, Sevilla and Tuluá. It initially covered more than 14,330 ha of coffee plantations. Out of these, 5,309 ha were already producing 4C compliant coffee. This verification system is outlined in detail in subsequent sections. Over the course of the next ten years, the Plan will be extended to other coffee growing regions in the country. In addition, some 60 million plantlets will be distributed as part of the coffee tree renovation efforts (Nestlé SA, 2012b; Nestlé Colombia, 2013b).

Partnerships

The worldwide coffee crisis in the 1990s mobilised non-profit certifying and labelling organisations to work to create a market fair for workers and sustainable for the environment. It also rang alarm bells amongst specialty roasters. They realised that consistently low international coffee prices compromised the capacity of many producers to sustain the labour-intensive cultivation, harvesting and processing required by high quality coffees. Unless something was done, these powerful actors would not be able to locate enough good quality coffee in the future. By then, coffee seems to have consolidated as the fire test for sustainability efforts, globalisation and trade for development initiatives. Already 15 years ago, the crop was one of the most internationally traded products. In the immediate aftermath of the crisis, a wide range of coffee actors first came together to jointly develop standards to address socio-economic and environmental concerns (Giovannucci and Ponte, 2005).

At the time, certified, verified and sustainable coffee emerged as a tangible way for consumers to be part of the solution to the problems surrounding its production (Linton, 2005). Sales are on the rise in this fast growing, true, but small segment. Expansion is mostly coming from parallel niche or upscale markets (Giovannucci and Ponte, 2005). The proliferation of certification schemes oftentimes sends the signal to consumers, as well as to producers, that there is a specific certification scheme to address one particular problem in coffee production. The environment is usually represented by organic coffee and the social stream is pursued Fair Trade. Castle calls them 'issue coffee' for consumers looking for 'anxiety-relieving certificates' (2001 in Linton, 2005). Without doubt, labels address some of the issues in coffee production.

Certifications have been recognised as a channel through which to reach the poor and stir large-scale changes in rural communities (SustainAbility et al., 2008). Yet, the interdependent and pervasive nature of many of the structural challenges in the crop's supply and value chain suggests problems can neither be compartmentalised nor tackled exclusively by civil groups. Large problems call for the mobilisation of large amounts of human and financial resources to bring solutions to scale. Private initiatives like the Nescafé Plan and the Nespresso AAA Sustainable Quality™ Program provide opportunities for a large number of suppliers to match higher quality standards. This transfers wealth downstream in supply chains as producers receive better incomes for their crops.

Large corporations, trading the bulk of coffee around the world, also have a vested interest in making coffee production sustainable. As the leading actors in many agrofood supply chains, retailers and manufacturers are becoming powerful forces of sustainability governance, both setting standards and driving social, economic and environmental change (Dauvergne and Lister 2012 in Rueda and Lambin, 2013). This certification and verification embrace has not been a controversy free process. Corporations have been accused of co-opting ethical trade initiatives and making the fair trade movement an adjunct of the conventional market (Jaffee, 2012). For many years, NGOs were averse to collaborating with the private sector.

Long reluctant to work with the private sector, some NGOs have begun partnering with businesses and large coffee purveyors. When they have done so, they have gained access to financial and human capital, market share and consumer influence. They can thus scale up good practices and accelerate the achievement of sustainability goals. In addition, mainstream coffee brands have the resources, marketing infrastructures and visibility that can educate consumers about sustainability issues (Bitzer et al., 2008; Linton, 2005). Once almost entirely antagonistic, large multinationals and NGOs have joint efforts to initiative large-scale change in the coffee industry.

Now that non-profit certifying and labelling organisations, industry-wide initiatives, multinational corporations and producers' associations have begun working together, the development of a more sustainable coffee market seems a goal within reach. As part of a partnership, these actors are able to use their comparative advantage to advance the sustainability agenda in the coffee trade. The aggregated impact of such efforts is finally promoting change in a supply chain that has not always worked in favour of farmers.

Easily identifiable by its logo, a green frog, the Rainforest Alliance is a non-governmental organisation based in New York. It has the mission to conserve biodiversity by promoting sustainable practices in agriculture, forestry, tourism and other industries. The group certifies, and seals, coffee produced under certain environmental standards (Ethical Coffee, n.d.). Since 1995, the Rainforest Alliance's sustainable agricultural programme trains and certifies small, medium and large farmers to improve their living conditions whilst encouraging them to protect their surrounding natural ecosystems and protect workers' rights (Rainforest Alliance, 2013). Rainforest Alliance-certified farms have to meet the social and environmental standards established by the Sustainable Agriculture Network. Rainforest Alliance-certified coffee products went mainstream in 2004, when the NGO first partnered up with Kraft foods and Procter & Gamble (Rueda and Lambin, 2013).

The Sustainable Agriculture Network (SAN) is one of the largest and oldest coalitions of non-governmental organisations working towards bettering commodity production via compliance with a series of social and environmental parameters for responsible farm management (SAN, 2010). It follows and advocates for the compliance with ten guiding principles, namely: (1) the existence of a social and environmental management system complying with national laws and SAN standards; (2) ecosystem conservation; (3) wildlife protection; (4) water conservation; (5) good working conditions for all employees; (6) occupational health; (7) community relations; (8) integrated crop management; (9) soil conservation; and (10) integrated waste management (SAN, 2010).

Founded in 2002 by 70 representatives from over 20 countries, the 4C Association has since then established a multi-stakeholder sustainable coffee platform that can unite coffee farmers, co-operatives, exporters, traders, importers, roasters, retailers, unions, public institutions, research centres, non-governmental organisations and individuals in mainstreaming sustainability practices prevailing in the sector. This business-to-business verification system assists farmers in working towards improving 28 economic, social and environmental principles for the sustainable production, processing and trading of coffee (4C Association 2010, 2012).

The 4C's 28 principles are grouped into the social, environmental and economic dimension. The social dimension benchmarks freedom of association and bargaining, equal rights and discrimination, the right to childhood and education, working conditions (contracts, compliance with national and international laws regarding working hours, wages, occupational health and safety, and equitable treatment for seasonal and piece rate workers), capacity and skill development, and living conditions and education.

Environmentally, the Code of Conduct offers a baseline for practices regarding conservation of biodiversity, use and handling of chemicals, soil conservation fertility and nutrient management, conservation of water resources, management of wastewater and wastes, energy savings and preferential use of renewable energy. In the economic dimension, the 4C system assesses shareholders' accessibility to commercial markets and market information, coffee quality monitoring, record keeping, transparent pricing mechanisms and traceability mechanisms (4C Association 2010, 2012).

As a founding member of 4C, Nescafé® has stressed the importance of farmer compliance with the Common Code for the Coffee Community as it benchmarks a series of actions all along the coffee value chain. This pre-competitive scheme does not provide 4C-compliant coffee labels for consumers to identify; instead, it encourages the entire sector to become more sustainable and transparent (Nestlé SA, 2014d; Muradian and Pelupessy, 2005). Industry players in the coffee sector compete for the same consumers and market shares. They are thus exposed to at least two latent supply risks: quality and scarcity. Pre-competitive practices, such as the ones pursued by the 4C code, build platforms where competitors set rules and make decisions jointly. Compliance is entirely voluntary.

The 4C Common Code, as an entry-level sustainability baseline standard, allows farmers to upgrade their practices and activities. Eventually, farmers are expected to comply with more demanding sustainability schemes such as the Sustainable Agricultural Standard (SAN), UTZ Certified or the Rainforest Alliance. The Code thus encourages farmers to discontinue unacceptable practices and move along the sustainability continuum (4C Association 2010, 2012).

Whilst it is true that the 4C Code assures a price premium for verified coffee, this higher price is only 2 to 3 cents per lb above the price for regular Colombian milds. The most advantageous element of the 4C verification is the technical assistance farmers can potentially gain access to in order to improve farming and processing methods. The expected higher yields and better quality can in turn result in higher farmer income. Moreover, the gradual discontinuity of harmful farming and processing methods can bring about environmental benefits, albeit much more modest than those attributed to the Rainforest Alliance Certification scheme (Nestlé SA, 2014d; Rainforest Alliance 2013). Since the mainstream market endorses 4C sustainability criteria, it is possible that with time terms of trade and planning may improve, risks may be reduced and farmers may gain access to a larger market segment. Complying with the sustainability scheme, however, requires moderately higher labour inputs (ITC, 2011).

The 4C Code was first implemented in Colombia in 2006 as part of a pilot project assessing the standards' applicability in the country. By 2008, the 4C Code of Conduct had been adopted by the Antioquia, Caldas, Risaralda, Santander and Valle del Cauca coffee districts. By 2015, the Nescafé Plan has aimed at directly procuring the 180,000 tonnes of green coffee it purchases every year from the Farmer Connect programme, which is fully in line with the 4C Code of Conduct. By 2020, the Plan has set the goal to source at least 90,000 tonnes of coffee following the sustainability principles developed by the Rainforest Alliance and SAN and distribute 220 million coffee plantlets (Nestlé SA, 2014a). With time, it is expected that involvement in this scheme can strengthen the economic and social resilience of the participating communities.

The 4C Association found coffee production in Colombia to be socially and economically advanced, with most of the gains to be achieved in terms of environmental protection (4C Association, 2010). This is probably because particular attention has been paid to the management of hazardous wastes and wastewater, as these are key action areas to boost sustainability in Colombia's coffee growing activities. Social initiatives have tended to focus on labour relations, employment conditions and occupational safety. Considerably less attention has been given to the social reality of coffee growing farmers themselves. To a certain extent, it has been implicitly assumed that productivity improvements will trickle down and result in higher standards of living as incomes rise. In reality, the average farmer has already been working the land for four decades, lives in poverty, attains only very low levels of formal education, possesses limited assets, does not have any sort of health insurance or pension scheme, and faces a narrow range of social and economic opportunities to lead a more prosperous life.

These partnerships with the Rainforest Alliance, SAN and the Common Code for the Coffee Community (4C) are key to allow farmers to gradually move from baseline to more stringent and inclusive sustainability standards. In doing so, they acquire the skills and resources they need to attain higher efficiency in coffee production. However, not everyone has welcomed the corporate engagement non-governmental organisations have established with big corporations such as Nestlé.

In fact, the Rainforest Alliance has had to defend its strategy to work with the private sector and fight back accusations that its credibility was being jeopardised and compromised. Companies, in turn, have been accused of 'green washing' their coffee supply chains by seeking association with reputable organisations offering sustainability certification schemes (Neilson and Pritchard, 2007). Multinationals and large coffee roasters and traders have been accused of establishing the 4C to keep their costs intact, blunt social criticism and salvage damaged public reputation. Code compliance, they argue, has been left to the limited devices of small and powerless producers (Menon, 2005 in Neilson and Pritchard, 2007).

Criticisms have also centred on the burden inflicted upon producer countries and farmers. During the May 2005 negotiations at the International Coffee Council Meeting to establish the 4C, producing countries stated they found the Code unjustly biased against the interests of producers and undermining to the sustainability of the coffee economy (ICC, 2005). Three years later, the 4C was seen much more positively. In 2008, Bitzer et al. gauged intersectoral partnerships for a sustainable coffee chain. They concluded that the 4C's broad membership made it one of the most promising initiatives to increase sustainability in the coffee chain.

Despite disapproval and sceptical responses, these public–private–social alliances have resulted in progressive transformations and improvements in farm management, environmentally responsible farming technologies and methods, as well as crop quality (Nestlé SA, 2012b; Nescafé Mexico, 2014). Together, they are actively addressing the multifaceted challenge of creating more responsible supply chains and the even more complex process of assisting coffee growers to secure their livelihoods and lead better lives (Rainforest Alliance, 2013; SAN, 2010; FNC, 2013a; 2013c).

Making the switch to sustainability is a tough challenge for farmers, who many times require high support. Local and international partners can thus work together to generate the sort of innovative dynamism that facilitates constant adaptation and learning to a continuously changing coffee sector. By pooling their expertise and resources, they fast track the acquisition of new production capabilities and upgrade existing skills. This makes farmers' sustainability learning curve steeper.

The record in Colombia

In Colombia, the Nescafé Plan is seeking to increase productivity, thereby helping small farmers to have higher yields and incomes. In 2012, over nine million disease-resistant coffee plants were distributed and 585 ha of coffee plantations renovated across the Valle del Cauca. Since 2011, the scheme has distributed more than 15.5 million disease-resistant plantlets and renovated 2,143 ha of coffee producing land. This strong focus on plantlet renovation has been mainly supported in municipalities neighbouring the Bugalagrande factory.

Beneficiaries see younger, more productive coffee estates and experience a decrease in violence, drug dealing activities and internal displacement. These are important farm and social changes and, despite not being entirely attributable to the activities undertaken under the umbrella of the Nescafé Plan, the scheme can be credited with fostering a peaceful and stable environment, social environment and more prosperous and promising economic landscape.

Given the popularity of this initiative and the positive impacts it has had on the region and its inhabitants, coffee growers outside of Nestlé's sourcing regions are actively seeking ways to take part in the brand's productivity and quality-enhancing actions. In 2013, benefits were further extended to reach 5,000 new farmers from Quindío and Risaralda, who received five million plantlets in the first year of implementation. In early 2014, the Nescafé Plan had enlarged its area of operation, assistance and support to an additional 335 ha in the municipalities of Belén de Umbria, Santuario, La Celia and Balboa in the department of Risaralda. As it has done elsewhere, the scheme is providing the technical and marketing conditions to improve productivity and quality, promote community development and give access to more profitable markets to around 5,685 coffee growers.

The first part of the Plan will seek to build farmers' capacities on best agricultural practices for all green coffee produced in those municipalities to comply with the 4C principles. Three agronomists have been hired to assist farmers new to the Nescafé Plan and two million rust-resistant plantlets delivered to continue the coffee tree renovation efforts in Risaralda. A total of 110 coffee growers will be trained in 4C standards to then guide farmers' groups in also reaching a minimum common sustainability and quality denominator (Nestlé Colombia 2014; La Tarde, 2014). The second component of the Plan will focus on boosting productivity and on delivering key inputs to increase production. Benefited coffee producers will also receive fertilisers, a crucial yet costly input to increase productivity and maintain those gains.

After only three years of operation, the Plan is already contributing to improvements in the economic and social resilience of communities that would otherwise be in the same precarious situation, similar to the other coffee growers in the country. In 2012 alone, Nescafé Dolce Gusto® bought 100,000 bags of sustainably grown Colombian coffee. Every year, the brand purchases of this type of coffee have increased by 16,000 bags. It can be confidently inferred that farmers selling their beans to the brand are facing improved terms of trade, productivity, quality and prices but the exact benefits of these transactions have not yet been quantified. These spillovers are described in more detail in Chapter 6.

More importantly, and firmly positioned at the core of the Plan, expert agronomists are improving the capacities and knowledge base of the farmers on sustainable coffee production practices. To September 2013, 2,477 coffee growers had received such training. In fact, this initiative has been highly popular among coffee growers. Farmers outside of Nestlé's sourcing regions are actively seeking ways they can take part in it. To transfer knowledge and know-how directly and as effectively as possible to almost 2,500 farmers every year, Nestlé is working with Colombia's long-established and experienced institutions, notably FNC (Figure 5.1).

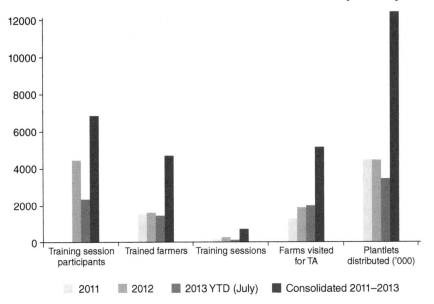

Figure 5.1 Benefits delivered by the Nescafé Plan
Source: Nestlé records

These efforts support the line of action followed by the FNC. Similar activities are being carried out in other countries where the Plan is being implemented, namely Mexico, Indonesia, the Philippines and Thailand. Based on observations in Colombia, the implementation and scope of the Nescafé Plan may have to be further fine-tuned and carefully localised to fit properly to the conditions, needs and profiles of each of the targeted communities. These long-term investments and efforts have the potential to radically and significantly improve the sustainability of the coffee production and marketing systems and the quality of coffee products available to consumers. On one end of the supply chain, coffee growers can improve their economic situation and have more resources to lead healthier and more prosperous lives. On the other end, consumers get to enjoy better quality, socially responsible and environmentally sustainable coffee products. One such example is Nescafé Gold Blend First Harvest, the result of the brand's joint efforts and work with the FNC.

Every year since 2011, the Nescafé Plan has distributed four million high-yield coffee trees in Colombia to rejuvenate farms. These new and enhanced plants contribute to overcome yield losses and heightened disease vulnerability caused by ageing and unkept coffee trees. The delivery of more than 15.5 million plants has been accompanied by free of charge technical assistance, investments, support, equipment, price premiums, improved marketing channels and the freedom for farmers to sell their coffee to whichever buyer they want.

These initial efforts to ensure the long-term supply of high quality, sustainably sourced coffee have already been taken to supermarket shelves. The rust-resistant, high-yielding coffee trees given to farmers in the Valle del Cauca region in late 2010 gave their first harvest in early 2014. These beans were used for the Nescafé Gold Blend First Harvest special edition, which contains coffee from farms supported by the Nescafé Plan. Launched on 30 April 2014, the blend is the result of the brand's commitment to responsible farming, production and consumption (Nescafé Mexico, 2014; Nestlé UK, 2014).

Creating shared value is without doubt a corporate policy and strategy, but its implementation in each factory, and along each supply chain, is very much context-conditioned and decentralised. It cannot be generically and automatically implemented, especially when it comes to large enterprises comprised of independent organisations that are expected to perform, grow and act as a unit. The size of a company such as Nestlé brings about challenges in terms of scale and in controlling all decisions across the entire enterprise and its numerous supply chains (Kytle and Ruggie, 2005).

The next section looks at the Nespresso AAA Sustainable Quality™ Program and how, similar to the Nescafé Plan, this private standard is functioning as a risk management and product differentiation marker. These sustainability-oriented schemes ensure that the two brands have sufficient control over production processes to guarantee that raw materials, of given quality levels and volumes, are available to purchase. They are also part of Nestlé's CSV framework. They conjointly generate growth for the company and benefits for the communities and environments producing coffee following improved agricultural practices.

6 Quality, productivity and sustainability

The AAA Sustainable Quality™ Program

Over a long period of years, and in an attempt to preserve its reputation as a high quality coffee producer, Nestlé has aimed at procuring the best available coffee around the world to fill its Nescafé® jars and later on Nespresso capsules. In the coffee market, these superior beans are labelled as AA. This classification, based solely on quality, conveys no information on whether the beans are being sustainably produced, the relationship between growers and the company buying them, the quality of life of the producers or the adoption of good agricultural and environmental practices.

To continue sourcing the highest quality coffees worldwide, Nespresso created an initiative that added quality and productivity dimensions to its commitment to promote sustainability in the coffee sector. This is reflected in the third A that appears on the labels of the brand's Grand Cru coffees (see Figure 6.1). As a private standard, the AAA Sustainable Quality™ Program plays two important functions: risk management and product differentiation (Alvarez and Wilding, 2008).

The AAA initiative complements sustainability principles with the highest quality criteria. Each A of the AAA Program represents the three pillars on which the Nespresso approach rests: quality, sustainability and productivity. Quality focuses on disseminating best practices, building capacities, enhancing traceability and improving coffee's sensory profile and increasing physical acceptance. To do this, the Program facilitates the acquisition and use of post-harvesting infrastructures and equipment (sun dryers, fermentation tanks and de-pulpers).

Sustainability is based on producer compliance with SAN principles as discussed in the previous chapter and once again in further sections. Productivity addresses tree renovation efforts and pruning cycles, land use optimisation, the creation of nurseries with high-yielding, disease-resistant varieties and the adequate use of fertilisers (Nestlé SA, 2011). Since its inception, together with the RA and SAN, the scheme has helped more than 60,000 coffee growers in sourcing countries all around the world to adopt sustainable quality farming practices.

Nespresso, in collaboration with the Rainforest Alliance, launched the AAA Sustainable Quality™ Program in November 2003 at Sintercafé. Every year,

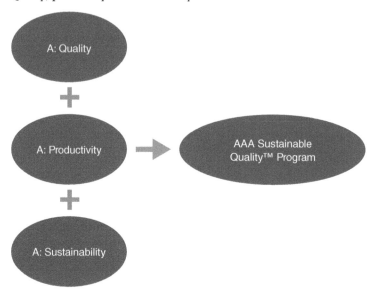

Figure 6.1 Beyond quality and productivity: sustainability

since 1987, Costa Rica has organised this one-week long industry initiative aimed at promoting the country's coffee sector (Sintercafé, 2014). The aspiration behind creating the AAA Program was to add the highest quality criteria to widely held sustainability principles. The timing was also right. The initiative was launched two years after real coffee prices were among the lowest in history. This coincided with Nespresso's exorbitant rates of growth. Securing coffee supply was thus becoming a more pressing concern. The AAA Program was developed at a time when farmers were facing insurmountable challenges to continue production. The collapse of international coffee prices in 2001 had devastating economic impacts that rippled through coffee dependent economies.

Subsistence farmers and families with small-scale farms initially reacted to the precipitous fall in household incomes through a combination of forgone production, increased migration and declining expenditures in education, health care and housing (Varangis et al., 2003 in Bacon et al., 2008). Lower investments in individual farms had ultimately translated into lower global output. Total coffee production was suffering and total yields decreasing. For its part, and during the immediate aftermath of the 2001 coffee crisis, the industry was undergoing important changes.

Activist organisations had launched extensive awareness campaigns to draw attention to the precarious economic situation many subsistence coffee farmers were consequently facing. At the same time, demand for traditional coffee products had begun to show signs of slowing down. Consumers were showing a strong preference for coffee products and retailers offering high quality drinks. In addition, the great success of Nespresso's capsules was pushing the company to grow above 30 per cent per annum. The interplay of these factors intensified competition among coffee traders who began facing escalating pressure to secure supplies of high quality beans. Farmers' precarious conditions were quickly perceived as latent risks potentially compromising long-term supplies of high quality raw materials for the rapidly rising single-serve coffee markets (Alvarez and Wilding, 2008).

The coffee crisis also mobilised NGOs to draw attention to growers' compromised livelihoods and other important industry players and large purveyors began developing sustainability projects. As the appetite for high quality coffee grew, roasters, civil associations, development agencies and coffee buyers enthusiastically began promoting the production of high quality and sustainable coffees. This led to the definition and execution of new global supply chain standards in developing countries and targeting small-scale producers (Perez-Aleman and Sandilands, 2008).

The resulting specialisation in coffee production was soon seen as a powerful and viable alternative available to farmers to overcome the crisis in the sector (Rainforest Alliance, 2013; Fairtrade International, 2013; UTZ, 2014; Alvarez and Wilding, 2008). New standards and certifications emerged as a mechanism to capture the economic value of complying with environmental, social and ethical parameters (Pierrot et al., 2010). Companies, on their part, moved quickly to design private standards to address risks of supplier failure. They mobilised their power and resources to ascertain, apply and audit their own sustainable agricultural codes, supplier guiding principles and quality assessment systems. In 2002, Starbucks became the first multinational company to announce the development of a preferred supplier system. In 2004, the Seattle-based coffee retailer launched the Coffee and Farmer Equity (C.A.F.E.), principles, through which it sources green coffee (Giovannucci and Ponte, 2005).

At this same time, NGOs like the Rainforest Alliance began working with corporations, like Nespresso. Forging these partnerships was a controversy-rigged effort. Businesses were attacked for trying to 'buy' legitimate social and environmental initiatives. NGOs working with agri-businesses were accused of 'selling off' to large corporations and of greenwashing industrial commodity agriculture. Many of them faced credibility and reputational risks and in a position of 'guilt by association' (Latham, 2012). Despite negative publicity, these alliances emerged as a way for brands to reduce transactions costs and access resources (Linton, 2005). For NGOs and firms trying to induce change in the coffee industry's operations, working together became a way of promoting change through more effective and larger-scale approaches (Manning et al., 2012; Alvarez and Wilding, 2008; Perez-Aleman and Sandilands, 2008). Partnerships, it turned out, reduce transactions costs, increase access to target groups and resources and link field realities with global market dynamics.

Former Nespresso CEO Gerhard Berssenbruegge was at the forefront of the launch of the AAA Program. It was put together as a strategy to simultaneously secure the supply of the highest quality coffee, reward farmers producing it and build stronger ties between consumers and farmers (PR Newswire, 2007). In

2004, one year after its launch, the initiative was extended to Mexico, Guatemala and Colombia.

At its initial stage, the Program was kept very flexible, open and experimental. Investments were small and scattered but proved successful in helping the brand secure long-term supplies of high quality coffee beans (Alvarez and Wilding, 2008). Goals have gradually changed but five considerations remain at the core: collaboration with institutional partners (the Coffee Growers Federation in Colombia), establishment of public–private partnerships, a focus on yields and quality (Qualitivity™), innovation in improving water management, and measurement and tracking of CSV performance (Nestlé Nespresso SA, 2008, 2013b).

The implementation of voluntary standards in the coffee sector, of which the AAA Program is one example, is regarded as one way to overcome many of the unsustainable aspects of coffee production (Panhuysen and Pierrot, 2014). The AAA Sustainable Quality™ Program is ultimately trying to induce positive changes in the operations of the coffee trade. As part of the initiative, coffee growers receive higher remuneration for the production of highest quality beans. This boosts improvements in their quality of life, as well as to their communities and the environment. They are also incorporated into a bottom-up path of continuous improvements that simultaneously raise quality and protect the environment. The inclusive initiative recognises the importance of developing interdependent and lasting company–customer, customer–farmer and company–farmer partnerships.

Commitments

Nespresso's business models rests on a series of assurances and objectives that cover every stage of coffee sourcing, production and retail. These are articulated in the Ecolaboration™ platform. This broader and integrated CSV framework integrates Nespresso's different sustainability efforts and commitments to responsibly manage the social and environmental impacts of its business activities.

The implementation of this scheme has strongly focused on fostering sustainable coffee farming in conjunction with partners and stakeholders such as the Rainforest Alliance for biodiversity conservation and the international coalition of conservation groups Sustainable Agriculture Network (SAN). The brand also works with the International Finance Corporation (IFC) – a body promoting sustainable private sector investment in developing countries – and the International Union for Conservation of Nature (IUCN). The Colombian Coffee Growers Federation (FNC) acts in representation of farmers. Other partners include industry players ECOM Agroindustrial Corporation, the Belgian green coffee and cocoa merchant Efico and the Colombian Coffee Growers Cooperatives Export Corporation Expocafé®. Joint initiatives have also been undertaken with TechnoServe, a non-profit organisation supporting rural entrepreneurship, as well as with the Central American Institute of Business Administration – INCAE, and the Swiss recycling group Thévenaz-Leduc.

As part of this framework, Nespresso made three key commitments to be attained by 2013. The first one focused on sourcing 80 per cent of the brand's coffee from the AAA Sustainable Quality™ Program. Further sections address how and when was this goal achieved. Already in 2010, the company was well on its track to reach it. Currently, around 84 per cent of the high quality coffee beans Nespresso secures to create its Grand Cru blends have been procured through the AAA initiative (Nestlé SA, 2014b).

The second Ecolaboration™ objective consisted of increasing to 75 per cent the capacity to recycle used capsules, a three-fold increase from 2009 levels. This was to be attained by putting in place capsule collection systems in the countries where the brand operates. By the end of 2013, Nespresso surpassed its objective, increasing its collection capacity to 80 per cent (Environmental Leader, 2012; The Green Organisation, 2012; Nestlé Nespresso SA, 2014c). The third commitment had to do with reducing by 20 per cent the carbon footprint of each cup of coffee. The next section looks at the measures taken to achieve the set cuts in CO_2 emissions.

In 2013, ten years after the AAA Program was first launched, the premium coffee brand formed the Nespresso Sustainability Advisory Board (NSAB) in support of its long-term strategy and initiatives towards improved environmental stewardship and farmer welfare. The Advisory Board is set to work on strengthening the brand's efforts to advance social and environmental sustainability goals. It is also to develop new schemes to create shared value for Nespresso AAA coffee producers (Nestlé SA, 2013d).

Meeting for the first time in 2013, the first NSAB put together a diverse group of people representing international NGOs, development organisations, producers associations, business schools and other stakeholders. To October 2014, its members include Harriet Lamb, CEO of Fairtrade International; Paul Rice, President and CEO of Fair Trade USA; Tensie Whelan, President of the Rainforest Alliance; Auret Van Heerden, President and CEO of the Fair Labor Association; Tristan Lecomte, co-founder of Pur Projet; Julia Marton-Lefèvre, CEO of the International Union for Conservation of Nature and Bruce McNamer, President and CEO of TechnoServe. Academia and the development sector are represented by Peter Bakker, President of the World Business Council for Sustainable Development; Polly Courtice, Director of the Cambridge Programme for Sustainability Leadership; and Lawrence Pratt, Director of the Latin American Center for Competitiveness and Sustainable Development at INCAE Business School. The participation of Luiz Genaro Muñoz, President of the Colombian Coffee Growers Federation, serves as further evidence of the importance Colombian coffee has for the brand. Lastly, Nespresso brand ambassador George Clooney has also joined the NSAB. His role consists of providing insights and recommendations to enhance the company's long-term sustainability strategy. Each year, the Board will focus on addressing specific sustainability issues, building multi-stakeholder partnerships and raising consumer and public awareness. Other goals include extending the CSV approach to intermediates and suppliers.

In its first edition, the NSAB focused on 'people' and the 'planet'. Members discussed the extension of the Nespresso AAA Sustainable Quality™ Program to Ethiopia, Kenya and South Sudan; the development of the AAA Farmer Future Program in Colombia; and the AAA Agro-forestry Program to encourage reforestation in and around coffee farming areas in Guatemala. The Board's second annual meeting introduced several major new initiatives. In 2014, a new ambitious strategy was set out to accelerate the company's sustainability efforts in coffee sourcing and social welfare; aluminium sourcing, use and disposal; and climate change resilience.

Brand growth has been accompanied by an expanding scope of publicly acquired commitments. In August 2014, the Nespresso Sustainability Advisory Board launched its most recent sustainability initiative: The Positive Cup. Directed at coffee sourcing and social welfare, the strategy is based on creating shared value and generating positive impacts for all stakeholders in the coffee value chain. Current investments are to be topped with a six-year long disbursement of a further CHF 500 million. A share of these resources has been allocated to establish a new Sustainable Development Fund, which will finance innovative and impactful projects. The Fund is also expected to channel resources from external partners into specific projects supporting the initiative's objectives. For instance, it will allocate funding to Nespresso's Program to revive the coffee sector in South Sudan (Nestlé Nespresso SA, 2014i).

These goals, to be attained by 2020, are undoubtedly ambitious: 100 per cent sustainably sourced coffee, 100 per cent sustainably managed aluminium and 100 per cent carbon insetting. The idea is also to succinctly transmit to consumers the brand's commitments to sustainability in coffee sourcing and consumption. Nespresso offers its Club Members a cup of coffee that can create a greater value for society and the environment (Nestlé Nespresso SA, 2014g).

Time and coffee farmers will tell how this investment performs in the future, especially as the brand's sustainability efforts are taken to new coffee producing countries in Africa and the number of farmers Nespresso delivers technical assistance to increases accordingly. To September 2014, the Rainforest Alliance–Nespresso partnership was working with over 60,000 farmers around the world to produce the highest quality beans the premium brand requires to satisfy growing demand in the portioned coffee segment.

Moving towards 100 per cent sustainable coffee sourcing

Embedded in these sustainability commitments were the brand's linkages to supplying farmers. They need to improve their long-term economic prospects, protect the environment and have a better standard of living. Nespresso has long-term requirements to secure supplies of excellent quality green coffee. This resulted in the co-development and implementation of the Nespresso AAA Sustainable Quality™ Program along with the Rainforest Alliance. In this way, social and environmental standards and the adoption of productivity-enhancing techniques are being jointly implemented. Put into action, these practices result in the sustainable production of consistently high-quality coffees whilst improving small farm productivity and conserving natural resources. In fact, they have positive impacts on farmers' livelihoods and income.

The Rainforest Alliance is a non-governmental organisation working towards conserving biodiversity. The focus of the AAA Sustainable Quality™ Program is to assist farmers to continuously improve the quality of coffee beans sold to the brand. With support from the brand, coffee producers work to improve on-farm environmental sustainability, increase economic profitability and further social development levels (Nestlé Nespresso SA, 2008; 2011c; 2012b). Nespresso made a commitment to source 80 per cent of its sustainability principles-compliant coffee by the end of 2013. This goal was reached in June 2013, six months ahead of the target date.

The year 2013 was crucial for the brand in terms of meeting and furthering its sustainability commitments. By the end of the year, which marked the tenth anniversary of the AAA Program, the Sustainable Quality™ initiative was supplying 84 per cent of Nespresso's purchases. This figure combines production originating from more than 29 clusters, extending over an area of 290,000 ha of AAA Program coffee plantations in Brazil, Colombia, Costa Rica, Ethiopia, Guatemala, India, Mexico and Nicaragua. The sustainable management scheme will be further extended to Africa, Asia and Latin America. With time, the brand aims at sourcing 100 per cent of its permanent range of Grand Cru coffees from sustainable farms (Nestlé SA, 2014b). This objective has been included as part of the Positive Cup, the brand's new sustainability strategy. Other goals include increasing the amount of Rainforest Alliance certified coffee to 50 per cent.

To reach these aims, the brand will have to rely on the investments and gains it has already achieved through the AAA Sustainable Quality™ Program. Continuity will be key, especially regarding assistance for farmers to achieve certification standards from Rainforest Alliance and Fairtrade. The brand will also have to work towards developing innovative solutions to challenges regarding farmer welfare, including the expansion of the AAA Farmer Future Program. This initiative, discussed in detail in further sections, is behind the creation of a retirement fund for coffee growers in Colombia (Nestlé Nespresso SA, 2014g).

From farm to landfill? What is happening to the capsules

Nespresso sales growth goes hand in hand with waste generation. The more capsules sold, the more pressing the need to put in place systems for adequate capsule disposal. This issue has received a great deal of attention from the brand itself as well as from environmentally conscious civil society groups.

Nespresso has taken a series of measures to address the growing waste challenge. It is starting by building sustainable supply chains for the aluminium used to manufacture the coffee capsules. In 2009, it launched the AluCycle™ Initiative by which the brand is able to significantly reduce its environmental footprint. Compared to sourcing primary aluminium, recycling it, a process that can be infinitely repeated, produces only 5 per cent of the carbon emissions.

Completing the recycling process is nevertheless rather complex and entails a great deal of consumer involvement. It also calls for collaboration with country-specific national environmental institutions and waste collection systems. Coordinated actions are required to arrange the logistics so that different actors and institutions are able to work together. Recycling poses additional, yet closely interlinked challenges: one of Club Member engagement in the recycling scheme and another one of convenience to recover used capsules.

Club Member engagement is an area of opportunity for the brand to reshape consumption behaviour and encourage customers to 'enjoy the good things in life', responsibly. That is, to be eco-committed. According to a 2012 customer segmentation study, 12 per cent of Nespresso's consumers fell under this category. Jointly, the AAA Sustainable Quality™ Program and Ecolaboration™ platform are ways in which the company is both responding to consumer demands for sustainable goods and mainstreaming sustainability in the production of its coffee systems (Nestlé Nespresso SA, 2014f). The initiatives are also examples of the extent to which consumers, through individual purchasing decisions and behaviours, can push for substantial, positive changes in the way supply and value chains are organised, governed and structured.

As a first option, and whenever available, Nespresso stresses the importance of working with national packaging recovery schemes. The earliest capsule collection system was put in place in the brand's home market, Switzerland, in 1991. There, capsule recovery has reached 99 per cent due to a set of recovery initiatives and the existence of 3,800 collection points. Swiss Club Members can dispose of their used capsules by taking them to Nespresso boutiques as well as to local and mobile collection centres and by making use of the free of charge, doorstep collection Recycling@Home scheme. This initiative was piloted in 2010, in London, and is now in place in six European markets (Belgium, France, Luxembourg, the Netherlands, Switzerland and the United Kingdom). The collection system allows consumers to return used capsules at the time a new capsule delivery is made (Nestlé SA, 2013c; Nestlé Nespresso SA, 2014c).

In other countries, capsules are collected through the Green Dot®, *Der Grüne Punkt*, waste recycling scheme. In Germany this option has been in place since 1993, in Sweden starting 2010 and in Finland from 2012. Through this European network of industry-funded systems for packaging recycling, consumers can use over 6,000 collection points to dispose their used coffee capsules (Nestlé Nespresso SA, 2013g, 2014c; Der Grüne Punkt, 2014). In France, since 2008, the brand has implemented a recovery system that now totals 5,000 collection points in boutiques, Mondial Relay® parcel shipment points, waste collection centres and, as mentioned before, on the consumers' doorstep. Additionally, in both the Swiss and French markets, Nespresso has developed an iPhone mobile phone application that helps Club Members find the most convenient way and closest collection point to return used capsules for recycling. In Luxembourg, Belgium and the Netherlands, Nespresso capsules can be collected at the collection and drop off points for the Kiala parcel delivery service. Alternatively, capsules can be handed out to be collected on the Club Members' doorstep (Nestlé Nespresso SA, 2014c).

Disposal concerns have emerged because such recycling alternatives are not available in most countries, especially emerging markets. In the absence of waste disposal and recycling systems, the brand has established its own recovery system. Collection points have been installed in some Nespresso boutiques. To the end of 2013, over 14,000 collection points had been facilitated in the following countries: Argentina, Australia, Austria, Belgium, Brazil, Canada, China, Czech Republic, France, Germany, Greece, Hong Kong, Hungary, Israel, Italy, Japan, Luxemburg, Mexico, Morocco, Netherlands, New Zealand, Norway, Poland, Portugal, Russia, Singapore, South Africa, South Korea, Spain, Sweden, Switzerland, Turkey, United Arab Emirates, United States and the United Kingdom (Nestlé Nespresso SA, 2013g). Capsule collection systems have also been customised for key business-to-business (B2B) customers such as France's Club Med, Bank of Austria, Erasmus in the Netherlands and Belgian Mobistar (Nestlé Nespresso SA, 2014c). At the industry-wide level, the coffee brand has also forged a series of partnerships to build a more sustainable aluminium value chain (The Green Organisation, 2012).

Under Nespresso's most recent sustainability strategy, the brand will work towards expanding its capacity to collect the totality of used aluminium capsules and increase recycling rates. In countries with well-developed and functioning recycling facilities – Belgium, France, Italy, Luxembourg, Netherlands, Switzerland and the UK – used capsules will be recycled to make new ones. The reality in developing countries is substantially different. Most of them do not have effective waste disposal management systems and it is likely capsules end up in local landfills. This is a problem the brand will have to find a solution to in the immediate future. Growth and success presents its own shortcomings, in this case, waste disposal.

At the industry level, the brand has harnessed its reputation and business muscle to reach mutually beneficial partnerships and accomplish shared goals. Created in 2012, the Aluminium Stewardship Initiative (ASI) is another example of a collective effort Nespresso has joined to reduce the environmental footprint of aluminium production, processing and recycling. This industry-wide initiative was co-founded and launched by Nespresso, the IUCN and 14 companies that form part of the aluminium value chain – including primary producers, transformers, convertors and commercial and consumer goods suppliers. The aim of this multi-stakeholder process is to establish a standard for sustainability and transparency in this metal's value chain by the end of 2014.

The standard is to be applicable for all stages of aluminium production and transformation, from bauxite mining, alumina refining, aluminium smelting, further processing and recycling. Together, ASI supporting companies are working towards creating a third party verified, 'responsible aluminium' product that measures and verifies progress and compliance in implementing sustainability practices. A first round of public consultations was held from 1 February to 29 March 2014 for industry, non-governmental organisations and academia to send their reviews to the ASI Standard's environmental, social and governance pillar (ASI, 2014). Under the Positive Cup strategy, Nespresso has made the commitment to make sure that 100 per cent of the virgin aluminium used for capsule manufacture is ASI compliant.

In 2009, and in order to reduce the environmental impact of its packaging, Nespresso co-founded the *Club du Recyclage des Emballages Légers en Aluminium et Acier* (CELAA – Club for Aluminium and Steel Light Packaging). The French group has the mission to increase recycling capabilities for small-scale aluminium and steel packaging such as coffee capsules, pet food containers and bottle tops. It thus devises and implements new recycling technology solutions, simplifies and harmonises waste categorisation norms and disseminates information regarding adequate waste disposal (Arpal, n.d.).

Carbon off and in setting

Nespresso focused its efforts on machine use and coffee growing, the areas of biggest environmental impact, to meet the 20 per cent carbon reduction target it set in 2009. Achieving that objective can only be the result of aggregate improvements across the machine, capsule and coffee production processes. For instance, a new generation of machines has been equipped with energy-saving automatic stand-by modes and power-off functions. Also, farmers in Guatemala have gained access to education modules on reducing farm greenhouse gas emissions and climate change preparedness. A smaller carbon footprint has also been attained with the installation of an innovative energy recovery system in the brand's production centre in Avenches, Switzerland. This technology has reduced the amount of energy needed to roast green coffee by 16 to 20 per cent. An overall 20.7 per cent reduction in the brand's carbon footprint can be attributed to these actions.

As part of the Positive Cup initiative, Nespresso has committed to achieve carbon neutrality and increase farm climate resilience. An extensive agroforestry programme will be rolled out to create healthier, more diverse, productive and profitable land systems. Participating farmers are being offered personalised technical assistance to improve farm ecosystems and implement agroforestry practices. They are also receiving plantlets free of charge and cash incentives to increase the number of trees planted. Increasing biomass on coffee farms compensates for emissions produced in the agricultural areas where trees are planted.

This approach, called insetting, aims at developing innovative projects that integrate long-term socio-environmental externalities into a company's core mission (Pur Project, 2014; Nestlé Nespresso SA, 2014h). By 2020, Nespresso, the Rainforest Alliance and Pur Project expect to have planted some ten million trees in AAA coffee regions worldwide. Pur Project is a collective of environmentally focused organisations collaborating with companies in incorporating climate issues into their businesses. The network works with companies in the regeneration and conservation of the ecosystems upon which they depend (Pur Project, 2014).

Added to this, Nespresso has made additional commitments to achieve a further 10 per cent reduction in overall carbon footprint. The results and impacts of this agroforestry programme will be monitored through the AAA database. Key performance indicators and geo-localisation of tress will be used to assess the effect of the initiative on coffee quality, socio-economic conditions and a series of environmental criteria, including soil and water quality, land use, biomass and pollination (Nestlé Nespresso SA, 2014g).

In collaboration with local coffee co-operatives, a first pilot started in 2013 in the Huehuetenango AAA cluster in Guatemala. Approximately CHF 150,00 have been invested in providing technical assistance to 150 farmers. Two on-the-ground technicians are carrying out this task. Furthermore, 50,000 timber and fruit trees were delivered in June 2014. On-farm vegetation will increase farm biodiversity and decrease carbon emissions. With time, as trees become productive, obtained harvests will constitute an additional source of income (Nestlé Nespresso SA, 2014g, 2014h).

Nespresso's 2020 vision once again highlights the importance of taking a collaborative approach to sustainability. Success largely depends on forging partnerships with actors who share sustainability goals and are ready and willing to mobilise their expertise to implement programmes at the local level. To expand the AAA Program to Africa, Nespresso will work with TechnoServe. To roll out the AAA Farmer Future Program, the brand has begun working with Fairtrade International and Colombian farmer co-operatives. Lastly, efforts to source coffee sustainably and implement agroforestry projects are being carried out with partners like the Rainforest Alliance and Pur Projet.

NGO–multinational corporation partnerships have contributed towards the dissemination and wide endorsement of values based on environmentally friendly and socially responsible coffee production and consumption. These alliances narrow down the knowledge and resource gaps separating farmers from domestic and global markets that reward higher quality raw materials with equivalent higher prices (Bitzer et al., 2008; Balch, 2011; Latham, 2012; Linton, 2005). In this way, sustainable livelihoods in small producing communities are linked with much larger-scale sustainable production and consumption systems (Hebebrand, 2011).

The sustainability continuum

Globally, Colombia plays an essential role for the brand as coffee supplier and a new market for Nespresso. The brand has announced the commencement of commercial operations in the country. A boutique will open in Bogotá later in 2014 (Nestlé SA, 2014a). Nespresso sources about a quarter of its green coffee from small farms in the Colombian regions of Cauca, Nariño, Caldas, Antioquia, Cundinamarca, Santander and Huila. These beans can be found in 80 per cent of the brand's Grands Crus, highlighting the importance Colombian coffee production has for Nespresso (Nestlé Nespresso SA, 2014d). Since 2006, the amount of coffee sourced from the seven Colombian AAA clusters has increased eightfold. This rate of growth is clearly the result of the Program's success and an indication of decreasing risks of supplier failure.

Coffee growers in this country constitute more than 65 per cent of the total beneficiaries participating in the AAA Sustainable Quality™ Program (Nestlé Nespresso SA, 2014e). These farmers receive personalised and farm-specific technical assistance to move towards higher on-farm environmental sustainability and productivity, economic profitability and better quality of life in terms of nutrition, education, household assets, etc. As a result of the AAA initiative, participating coffee growers have more productive farms and obtain better yields than their counterparts, and thereby receive higher prices for the production of highest quality beans. Better incomes contribute to improvements in their quality of life, which also has positive socio-economic impacts on their communities and the environment. By upgrading and improving coffee's productive infrastructure and guaranteeing good-quality coffee beans and products have access to premium markets, the brand is mitigating supply chain risks. In the past, social and market uncertainty faced by the growers have intensified the risks of producer failure.

To produce the differentiated coffee beans that Nespresso uses, farmers need to develop more capabilities and make greater and regular new investments. As part of this process, they follow a farm management plan according to the Tool for the Assessment of Sustainable Quality, TASQ™. The plan has been designed as a bottom-up path to continuous improvement that simultaneously raises quality and farm incomes, and protects the environment. The self-assessment tool was created in 2004 with the support of SalvaNatura, the Natura Foundation and Rainforest Alliance, three SAN member organisations. Depending on their capacity to incrementally introduce changes, producers work towards improving compliance in four key aspects: quality, economics as well as social and environmental issues (Sustainable Coffee Project, 2014).

Farms are the basis of sustainable quality and the way they are managed determines the quality of the final product. This highlights the significance of engaging and training farmers in assessing their own practices. To assist them in gauging current practices, Nespresso uses a short, illustrated guidebook. This easy to use and fill booklet contains self-explanatory images that help farmers guide their answers. Once filled in, it provides general information about the farm, including the plot's name, location, number of hectares under coffee production, number of coffee trees, average tree age, use of fertilisers, tree renovation cycles, total coffee production, tree varieties and number of workers employed. Subsequent sections capture quality-oriented practices followed in coffee processing, namely at the time of harvesting, pulping, removing mucilage, fermenting, drying and storing dry parchment coffee. Farmers are also requested to record hygiene practices and the quality and amount of water used per kg of processed coffee (Nestlé Nespresso SA, 2012b). What the booklet does is to set a baseline for each individual farm against which to assess future progress in farm management practices.

To establish the level of sustainability already existing on a given farm, coffee growers must state whether they take any measures to facilitate coffee traceability; protect natural areas and water sources; protect biodiversity and manage wastewater; offer adequate labour conditions; and avoid the use of forbidden chemicals. They also submit information about their level of community engagement (Nestlé Nespresso SA, 2012b). This exercise helps both farmers and Nespresso make a farm diagnosis. With this information, it is possible to identify the priority areas in which growers require technical assistance. As much as possible, specific support is offered for farmers to attain the quality the brand requires by advancing farm practices.

Reaching sustainability and attaining exceptional coffee quality are the result of gradual processes of continuous and sustained improvement. TASQ™ recognises this and allows farms with deficient practices to join the AAA Program. Farmers' progression towards certification remains limited until suboptimal, deficient or harmful practices are substantially changed or discontinued. Coffee growers move along a 4-Step sustainability continuum (from Step 0 to Step 3) depending on their own capacity to comply with different criteria. Step 0 is an AAA entry-level stage prior to compliance with 32 compulsory criteria that allow farmers to move to Step 1. The baseline stage assesses whether farms pose a high social, economic and environmental risk depending on the practices under which they are managed.

Deficient practices include the lack of systems to avoid AAA coffee getting mixed with other qualities as well as documented transactions for a farm's coffee production and beans that qualify as AAA. Other critical parameters have to do with the absence of systems and practices for the identification, protection and conservation of the natural ecosystem, water sources, etc. The introduction of genetically modified crops is also to be avoided. Social risks are evaluated through indicators capturing the adoption of appropriate labour practices and respect for national labour laws regarding child labour, minimum wages, working hours, the use of health and safety equipment, as well as the use of prohibited substances. As a baseline stage, it also establishes a minimum common denominator for community involvement (Nestlé Nespresso SA, 2009b). Entry barriers to the sustainability scheme are kept initially low. Quality and sustainability demands are gradually introduced. Additionally, farmers are given support and assistance to resolve specific criteria in which they fall short. However, continuous failure to comply with sustainability parameters results in exclusion from the AAA Program.

Coffee growers adopt better farming practices depending on the Step they are in. They gradually move from deficient to basic, emerging and advanced performance levels. As compliance with SAN sustainability principles increases, farmers become eligible to receive additional premiums for their coffee. Every year, *NaturaCert* trained auditors verify farm TASQ™ compliance. Auditors measure and record progress in over 300 criteria that are both part of SAN principles included in the Tool's 4-Step sustainability continuum. A Natura Foundation initiative, *NaturaCert* has been designed to offer certification and verification services for sustainable standards.

Grouped into ten principles, sustainability objectives focus on coffee management systems (planning, training, monitoring and follow-up, coffee quality and sustainability traceability, energy efficiency and economic management); environmental conservation; wildlife protection and water conservation (including the management of wastewater). Other principles include fair and good working conditions; occupational health and safety; community relations; integrated crop management (plague control and the use of agrochemicals as well as of restricted and prohibited products); soil management and conservation; and integrated waste management (Nestlé Nespresso SA, 2009b, 2009c).

The TASQ™ system is homologated with SAN standards. This enables farmers in the Program to identify the areas to be strengthened in order to qualify for higher sustainability levels. Coffee growers selling their beans to Nespresso's AAA Program should also strive to comply with RA's basic principles and eventually obtain this certification. As a result of continuous improvement in coffee production by farmers, by the end of 2013, more than 30 per cent of Nespresso's global coffee was RA certified. It should be stressed that to belong to the Sustainable Quality™ Program, farmers do not have to incur any additional costs nor pay any fee either for the RA certification or AAA verification. In fact, since 2009, the brand has covered the costs of the first year of RA certification audits.

What farmers are actually required to do is to follow a comprehensive social and environmental management system that includes soil management and preservation, integrated waste management and the establishment of community links. Compliance with these sustainability principles exemplifies progress in the AAA Program and opens up opportunities to apply for the Rainforest Alliance certification (Nestlé Nespresso SA, 2013b, 2013f). As farmers obtain this seal of approval, they are able to secure additional income from additional premiums paid for coffee produced following environmentally sustainable practices in at least two ways.

First, Nespresso pays a specific RA premium on top of the quality premium paid for AAA beans. Second, farmers can obtain RA premiums for the coffee they do not sell to Nespresso. The brand is simultaneously mitigating supply chain risks and market uncertainty by upgrading and improving coffee's productive infrastructure and guaranteeing good-quality coffee beans have access to premium markets. The inclusive initiative thus recognises the importance of developing interdependent and lasting company–customer, customer-farmer and company-farmer partnerships for companies to grow and farmers to stay in the coffee trade.

The record

All these sustainability and quality-oriented schemes have created additional value for more than 40,000 farmers. In 2010 alone, growers obtained more than USD 19.5 million in additional income (FNC, 2011a). Between 2006 and 2013, Nespresso AAA purchases in Colombia multiplied more than eightfold. This extraordinary rate of growth is partially explained by Nespresso's initial expansion from one geographical cluster in partnership with one supplier (Expocafé®) and 5,000 coffee growers in 2005 to operating in three clusters, grouping 35,000 farmers and working with three suppliers – FNC, Expocafé® and SKN Caribecafé. Clusters are geographical regions where coffee procured meets Nespresso's quality and aroma standards; beans have the same global characteristics in terms of altitude, soil and varieties; production and quality are stable; full traceability can be assured; and there are relationships between producers, co-operatives, suppliers and associations.

To March 2014, the Nespresso AAA Sustainable Quality™ Program had grown to encompass seven clusters in Colombia: Antioquia, Caldas, Cauca, Cundinamarca, Huila, Nariño and Santander. Colombian coffee sales to Nespresso have increased more than eightfold since 2006 (FNC, 2014a). Such growth rates indicate the Program's level of success in securing high-quality coffee, and the economic attractiveness of the initiative to the farmers. At the same time, higher and better-paid coffee purchases have translated in higher incomes for the AAA coffee growers and their families. They also offer an incentive for farmers to continuously improve the quality of their beans and to look at coffee production less as a subsistence-level family endeavour and much more as a profitable business. On average, Nespresso premiums have consistently been 30 to 40 per cent above the New York stock market price paid for standard coffee and 10 to 15 per cent above the price for coffees of similar quality.

It should be noted that these sustainability and quality-oriented initiatives have made significant contributions to Colombia's coffee sector and the farmers selling beans under premium paying, sustainability-oriented initiatives. The country sold 1.2 million bags of speciality coffee in 2013, an increase from a record high of 1.03 bags in 2011. The FNC has estimated that 29 per cent of the country's total producers, some 162,873 farmers, are selling their coffee to premium and differentiated markets (Market Wired, 2013). Access to specialty markets allows farmers to obtain higher prices for their crops and offers them better protection against price shocks and volatility (FNC, 2014b).

Regarding environmental sustainability, together, Nespresso, FNC and a group of development and social partners have helped, for example, take decisive measures to anticipate and counteract bottlenecks in the availability and management of natural resources, mainly water. Between 2008 and to June 2013, more than 29,000 wet milling and sun drying installations (including 15,000 individual water treatment units), 8,000 water management solutions, 5,000 fermentation tanks were installed in different farms and upgraded to better manage water resources (Nestlé Nespresso SA, 2013e).

These actions have also further contributed to curb down carbon emissions. Methane (CH_4) and nitrous oxide (N_2O) emissions from pulp piles have been reduced (Ernst Basier + Partner, n.d.). Just four years after the AAA Sustainable

Quality™ Program was developed, Nespresso received the Rainforest Alliance's Corporate Green Globe Award on 16 May 2007. This recognition was given to the brand for showing exemplary and extraordinary dedication, innovation and commitment to furthering environmental sustainability. Nevertheless, since individual water treatment facilities are rather costly to purchase, install and maintain, it has proven key to fully assess how effective they may be over the long term.

Starting in December 2012, and for 12 months, *Fundación Natura* and the Institute for Research and Development in Water, Sanitation and Water Conservation (CINARA) conducted a review exercise and diagnosis of treatment systems Nespresso installed in Caldas and Huila. Cenicafé (Colombia's National Coffee Research Centre) is currently analysing the test results. Once completed, the evaluation will define the technical feasibility, functionality, and operation and maintenance challenges of the different wastewater treatment systems for wet coffee installed in that region. Based on the test results and reports prepared by CINARA, *Fundación Natura* and Cenicafé, Nespresso will validate and/or adjust its current strategy to improve farmers' access to wastewater treatment systems.

Water conservation is an important environmental goal. It also has verification and certification implications. Small farmers face considerable financial obstacles in acquiring the wet coffee processing wastewater treatment systems they need to reduce their water footprint as well as to comply with the applicable Colombian Technical Regulation for the Water and Sanitation Sector and SAN's fourth principle on water conservation. Meeting the requirements of the Sustainable Alliance Network is fundamental for farmers to be able to obtain and retain Rainforest Alliance certification. Alongside projects targeting and assisting individual farms to acquire wastewater treatment systems, collective and larger-scale initiatives will add up to more sizeable gains. Water resources can be more efficiently used, water management improved and pollution reduced.

Picking friends

Collective solutions have proven to be a more cost-effective way to tackle the two main priorities of the AAA Program in Colombia: harnessing the efforts of government bodies and local partners to halt and reverse the decline in productivity of recent years; and significantly improving the management of residual waters. Most small coffee producers lack the necessary resources, capacity and knowledge to cover the sunk costs required to make investments in water infrastructure, be it for domestic uses or coffee processing. These challenges are best tackled through collective programmes. Joint initiatives also allow different actors, from different sectors, to work together. Each one harnesses its expertise, resources, critical mass and organisational comparative advantage.

For instance, the FNC, Nestlé, Nespresso, the Colombian Ministry of Agriculture and Rural Development, Cenicafé and Alterra-Wageningen UR in the Netherlands and the Dutch Ministry of Foreign Affairs are exploring ways of working together to alleviate poverty in coffee growing regions by helping farmers become increasingly more self-reliant. These partners have come together to start an Intelligent Water Management System. They have proposed a five-year long project to establish basic environmental, social and productive conditions and address water pollution and climate change challenges in the regions where the Nescafé Plan and the AAA Program are operating (Wageningen UR, 2013).

Another example of successful collaboration between the private, development, social and public sector has helped to strengthen key processes to assure good quality coffee. An extensive consortium of partners, including Nespresso AAA, Agricultural Cooperative Development International/ Volunteers in Overseas Cooperative Assistance (ACDI/VOCA), USAID, the Coffee Growers' Cooperative in the Andes and the coffee exporter co-operative Expocafé® came together to co-finance a Central Mill in Jardín, in the department of Antioquia. The facilities were built in less than six months. Doors opened on 10 December 2010. Besides the wet mill, where cherries are fermented and washed, the facilities include a quality testing laboratory, an integrated solar drying installation and a waste management system.

Run by a co-operative, this environment and local ecosystem-oriented project has simultaneously improved bean quality and productivity for 170 farmers in Jardín and improved the environmental sustainability in the region (FAO, 2013; Nestlé Nespresso SA, 2013c, 2013d). This has proven to be an example of successful collaboration between the private, development, social and public sector as it has helped to strengthen key processes to assure good quality coffee.

Coffee growers take, or send by lorry, the cherries to the Mill, where the bags are received and weighed, and the information of the producing farmer entered into the local co-operative's database. Daily coffee prices and premiums are announced on a board at the delivery point. There, a sample of each coffee batch is taken to estimate the amount of over-ripe coffee and other defects. Sun-dried (natural) coffee is screened to identify the following defects: black, sour, rancid, parchment, broken green and insect damaged beans; number of cherries, husks and shells; and the presence of stones, earth clods and sticks (ITC, 2011).

The premiums farmers receive for good quality beans are based on the expected volume of AAA coffee that will result from the milling and the drying processes. Since the Mill has been in operation, farmers have produced higher quality coffee. The volume of AAA beans sold has doubled. As a result, farmers are securing more revenue from premiums and progressively obtaining net incomes up to 30 per cent higher (Nestlé Nespresso SA, 2013c; FAO, 2013). The Mill has also provided an instant post-harvest cash flow. Coffee growers are paid when they take the cherries to the Mill as opposed to when the coffee is processed. Additionally, local economies reap additional benefits as any profits generated at the Mill are shared among the co-operative's members at the end of the season.

Milling and drying are labour and water-intensive processes. They are even more so for average coffee growers, who usually own farms as small as 2.5 ha and rely solely on household labour. The Central Mill has redistributed coffee growers' traditional workload as farmers are offered the possibility of delegating these tasks. Employees at the Jardín Central Mill, all of them people from the region, are now in charge of performing the process farmers have traditionally followed to transform cherries into coffee beans. As a result, more than four hours of labour and water intensive tasks are saved during the harvest. Many farmers are investing the time saved in growing fruits that serve as additional sources of family food and income. As a positive, although not yet quantified externality, the creation of milling infrastructure has improved local food security.

The difference between a premium quality bean and a low quality one is determined during the milling and drying processes. If the quality of parchment coffee is compromised, so are the premiums paid for the beans. Processing centralisation decreases the likelihood of damaging the coffee cherries and allows farmers lacking the adequate infrastructure or resources to carry out the drying process to sell cherry coffee. Moreover, the taste and aroma that give coffee the Grand Cru profile is maintained by sun-drying the cherries in the Mill's 3,000 square metre installation. All these actions have brought down rejection rates, which translate into higher AAA quality coffee sales and incomes. The wet milling centre has effectively created shared value through infrastructure (Porter and Kramer, 2011).

At the co-operative level, parchment rejections due to inadequate physical characteristics have gone from 0.4 lbs to 0.1 lbs per bag. Laboratory rejections carried out by Expocafé® have fallen from 2.1 lbs to virtually zero. Dry mill losses have more than halved, from 1.5 lbs to 0.7 lbs. In total, rejection rates have gone from almost 50 per cent to almost zero. For Nespresso, centralised coffee processing has led to more efficient coffee supply. The amount of parchment required to obtain 1 lb of AAA green coffee has been halved from 6 to 3 lbs (Nestlé Nespresso SA, 2013b, 2013d).

Environmentally, by processing coffee at the Mill, on-farm water consumption has been reduced by 63 per cent. Yearly, each participating farm is saving an estimated 27,000 litres of water. For the average AAA farmer, on-farm coffee processing takes 25 litres of water per kg of coffee. In comparison, by taking the cherries to the Mill in Jardín, the process only consumes 11 litres. In addition, the Mill has also taken to treat highly contaminated wastewater, previously discharged untreated to water bodies, or left for disposal by surface runoff.

Following the economic and environmental successes that have resulted from the centralisation of milling and drying processes at the Central Mill in Jardín, a second facility will be built in the department of Huila. With more such facilities, and intensified activities in existing ones, the amount of waste is bound to increase. At the time of the study, a pilot initiative was using pulp residues as compost. It remains to be seen if it can be implemented at a large scale or if other residues can be used in any other way. For instance, pulp residues could generate energy through biogas or ethanol production, combustion, pyrolysis (the thermochemical decomposition of organic material), gasification, briquetting, and pelletising (Ernst Basler + Partner, n.d.).

Through the Nescafé Plan, Nescafé® and Nespresso have given proof of their level of commitment to environmental sustainability, economic profitability and social development. The two brands, one for the mainstream market and the other for the premium end, have proven to be ideal partners for the FNC as they boost the image of Colombian Mild Arabicas all around the world (Nestlé Colombia, 2014). The Nestlé–FNC alliance has produced one of Nespresso's most successful coffee varieties. When the AAA Program started in Cauca in 2005 it did so with 574 farms. One year later, in 2006, Nariño joined the AAA Program, both clusters adding up to 1,961 farms. In 2009, enough coffee growers had joined the AAA Program for Nespresso to launch a permanent Grand Cru, Rosabaya®, sourced mostly from this region.

As one of Nespresso's three pure origin coffees, Rosabaya® is the product of the close partnership that exists between the brand and the FNC. It is also 100 per cent AAA coffee. Up to 2103, and with the support of 94 FNC agronomists, supported by Nespresso, some 30,000 families have joined the initiative. In 2012, the United Nations Environment Programme (UNEP) recognised this partnership, and the programme Nespresso and the FNC put in place in Cauca and Nariño was considered to be an example of a sustainable business model implementing good environmental, social and economic practices. The brand and the Coffee Growers Federation were recognised for their joint activities contributing to the Green Economy movement (FNC, 2012a).

Assessing sustainability

At the same time that the company is securing enough coffee to sustain its double-digit annual growth rate, producers are selling their fine AAA coffee at premium prices. In addition to net incomes being 17 per cent higher, social conditions for benefited farmers are 22.6 per cent better, environments 52 per cent greener and economic prosperity 41 per cent above that of mainstream farmers. This is an example of how creating shared value is all about sharing success at all stakeholder levels.

The Centre for Regional Entrepreneurial and Coffee Studies (CRECE) conducted a survey amongst 1,222 AAA farmers in all Nespresso AAA clusters using the COSA™ methodology. The COSA™ methodology was developed by the Committee on Sustainability Assessment (COSA™), a consortium of non-profit organisations and think tanks studying the implementation of sustainability programmes. The methodology analyses the economic, environmental and social aspects involved in sustainable commodity production. The Committee seeks to develop tools that generate statistically significant, globally comparable data. Quantitative indicators deliver important data to better understand, manage and accelerate sustainability (COSA, 2013).

Carried out from 2009 to 2012, this assessment tried to capture the impacts that the Nespresso AAA Sustainable Quality™ Program have had on the quality of life of the Colombian coffee growers supplying beans to the company between 2009 and 2011. The study also sought to determine the difference the AAA scheme is making amongst participating communities in terms of environmental, social and economic sustainability.

The CRECE study built a sustainability-monitoring index based on social, economic and environmental indices. The social index captures working practices, living conditions, occupational health and safety conditions and social perceptions held by coffee farmers. The economic index reflected farmers' market knowledge, land productivity, crop yields, production costs, income and perception of business opportunities. The environmental index comprises the adoption of good agricultural practices, soil and water conservation measures and agrochemical handling.

The initial 2009 study provided quantitative evidence of the benefits AAA Advanced (SAN Compliant) and AAA farmers have gained since they joined the Sustainable Quality™ initiative and obtained the Rainforest Alliance certification as part of the AAA quality seal. This evaluation showed that the surveyed SAN-compliant, advanced AAA farms and AAA farms performed better than non-AAA farms in social, environmental and economic terms.

Additionally, a subsequent follow-up evaluation undertaken in 2011 indicates that the AAA Advanced and AAA farmers continuously improved in each one of the sustainability pillars. In the 2011 study, the largest gaps were recorded in terms of environmental practices as AAA farms scored 52.1 per cent higher than non-AAA farms. AAA Advanced farmers obtained even higher scores, with 79 points in terms of improved environmental conditions. Economic differences came second, with AAA Advanced and AAA farms obtaining 66 per cent and 41 per cent more points than the control group in the economic index respectively. Finally, in the social index, AAA Advanced and AAA farms recorded a 26.4 per cent and 22.6 per cent difference with the non-AAA control group.

Much importance has been awarded to economic and environmental aspects, as clearly reflected in the wide differentials between AAA Advanced, AAA and non-AAA farms for these two indices. AAA Advanced and AAA farms have been obtaining significantly higher results than control group farms, which is a reflection of a widening of the gap in how agricultural practices are improving in the economic and environmental fields. For instance, in 2009, AAA coffee producers obtained 68 points in the environmental index, 38.8 per cent higher than the 49 points scored by the control group.

By 2011, the largest gaps between AAA Advanced farms and conventional farmers were recorded in terms of environmental practices. AAA Advanced farms scored 79 points and AAA farms 73. In contrast, growers outside the programme actually saw a slight deterioration in environmental practices as the index marginally fell from 49 to 48. That year, AAA farms scored 52.1 per cent higher than non-AAA farms. For instance, 87 per cent of AAA farms have recycling programmes as opposed to average farms, where similar schemes are only in place in 43 per cent of them. Similarly, soil conservation practices are more widely adopted and carried out by AAA farmers.

Economically, the control group was worse off in 2011 than in 2009. Even if the index only fell marginally from 40 to 39 points, it does illustrate the many challenges the average Colombian coffee growers are facing to remain competitive. AAA farmers have seen a 6-point improvement in their economic indicators, from 49 to 55 points, which has also considerably increased the differentials with non–AAA farms from 22.5 per cent in 2009 to 41 per cent in 2011. Differentials are even wider between mainstream coffee growers and AAA Advanced farmers. This latter group obtained 59 points in terms of economic conditions, which was the highest score among all farmers' groups and was 4 points above the score obtained by AAA farmers.

Nespresso AAA farmers are better prepared regarding marketing issues, farm management and traceability. They have also tried to find new customers since their perceptions of the farm's economic situation and business opportunities have improved. On these coffee farms, rust affects less than 10 per cent of the trees, and farmers follow good fertilisation techniques that are recommended to them and keep records of the applications. This group is aware of domestic and international coffee prices and also has a higher net income.

Healthier, younger trees are the result of the implementation of the 'Tree of Life' (*Árbol de Vida*) initiative, aimed at increasing the number of improved coffee trees by planting two million rust-resistant coffee trees. Along with crop-management and output-enhancing efforts, renovated trees have significantly bolstered productivity. For instance, in Pacora, Caldas, crop management systems, soil analyses and the adequate use of fertilisers have contributed to a productivity increase between of 6 and 30 per cent in the benefited farms. Individual progress in yields and quality, the result of the successful implementation of the different Nestlé and Nespresso supported projects and a sense of ownership amongst farmers, has doubled the volume of coffee matching the Nespresso quality criteria. Nonetheless, Program loyalty is greatly driven by the premiums farmers receive for their coffee. This is not surprising since the economic situation producers generally face has been particularly dire for almost 20 years.

Studies looking at the impacts of certification schemes at the community level have thereby been primarily concerned with the economic advantages accrued to participating farmers. Certifications, verifications and sustainability-directed initiatives attract the participation of farmers because they pay higher prices and premiums for sold beans. As it has been discussed earlier, higher incomes can unleash a series of positive social benefits if reinvested in education, health, housing, nutrition and training. However, there is evidence that there are more benefits to sustainability than higher premiums and incomes. There is a series of additional benefits – for instance higher levels of relational capital, knowledge spillovers and technological diffusion – of which not enough is known (Giovannucci and Ponte, 2005).

Mindful of the fact that price is the main reason behind farmers' preference to sell to a given trader, a study looking at the impacts of certification on Colombian small-scale coffee growers suggests there is more to the story than just premiums. Rueda and Lambin (2013) conducted an evaluation of the Rainforest Alliance certification scheme in the coffee growing region of Santander. They tested whether such an initiative benefits farmers in non-monetary ways. They were especially interested to see if the RA certification strengthened farmers' responsiveness to market changes. Their study also assessed whether the seal fosters the conservation of ecosystems and what the main elements incentivising farmers to join and remain part of the programme are. Their findings are particularly relevant and interesting to the topics discussed in this book. For instance, Nespresso has been collaborating with the Rainforest Alliance in the different clusters from where it sources its coffee.

The certification programme began in Santander in 2002, a year when international coffee prices were at ominously low levels, averaging a mere 40 cents per pound of green coffee. Not surprisingly, the promise of receiving a premium paid for certified coffee persuaded many farmers to follow the agricultural practices endorsed by the Rainforest Alliance. In 2002, premiums reached 17 cents per pound of coffee – a gain 40 per cent above the price producers were receiving for mainstream beans. As international prices for Arabica beans rose and the availability of RA-certified coffee increased, the importance of premiums decreased as a total share of income. In 2013, Rueda and Lambin estimated that premiums represented an additional income of 3 per cent or less.

Farmers must then value gains other than just price differentials. Their continuous participation in this certification initiative highlights the importance of non-pecuniary benefits accrued to them. Rueda and Lambin (2013) listed access to information, technologies, social networks, resources and upgraded market outlets offering more stable prices. Ultimately, new knowledge has translated into more resilient social–ecological systems and responses to changes in the global coffee sector.

The implementation of both the Nespresso AAA Program and Nescafé Plan has also brought about additional benefits that have not been systematically recorded. This is perhaps attributable to the fact that these positive developments include attitudinal changes that are mostly related to, and do not directly cause, the creation of more landscape to introduce sustainability oriented practices. Yet, these benefits were recurrently mentioned during formal and informal conversations held with farmers, technical advisors, co-operative employees and people working in the coffee sector. These positive impacts have had an imprint at the individual, household, community and institutional level.

The sustained increase in productivity and quality suggest growers are indeed adopting the knowledge and practices that satisfy the standards set in the AAA Program. As capacities are built, engrained and replicated, farmers are acquiring more and better tools to make coffee production a more profitable activity. In terms of credit, however, the position of Nespresso coffee growers needs to be closely assessed. According to CRECE (2011), in 2008 only 14 per cent of Nespresso coffee growers had access to credit, which was the lowest rate amongst conventional and certified groups. Such low percentages can be due to low credit demand due to the additional assistance farmers receive as part of the Sustainable Quality™ Program and from development partners such as the United States Agency for International Development (USAID).

Regular farmers can only commercialise 23 per cent of their crop in the speciality market. This rate is much higher amongst AAA farmers, who are able to sell 66 per cent of their coffee in the high-quality bracket and thus receive the premiums paid above rates received for regular coffee. They also receive 40 per cent more technical support than the control farmers. This encourages them to plant higher numbers of rust-resistant varieties and renovate their plantations at faster and higher rates.

Intensified support is generating technological and economic differences between AAA and more traditional farmers. This explains why AAA producers have suffered fewer productivity losses. Consequently, AAA coffee growers are increasingly selling more coffee to the company and are establishing stronger loyalty ties with Nespresso. Overall, AAA farmers have net incomes up to 46 per cent higher than those of non-AAA producers.

Nespresso AAA coffee growers were reported to be at least 30 per cent more productive than farmers in the control group and are able to sell 78 per cent of their coffee as high-quality coffee, receiving the premiums paid for these better quality beans. Despite such wide productivity gaps, further gains can be secured to, at the very least, match the high productivity levels attained by farmers selling under the Fairtrade label. These growers were 47 per cent more productive than the average farmer. In Huila and Nariño, Nespresso AAA farmers obtained yield 40 per cent above those of mainstream coffee producers. Productivity and yields differentials meant that gross margins for Nespresso farmers were 7.2 times above those obtained by the average.

The CRECE study shows the pivotal role training on good agricultural practices play in making farmers more competitive. Despite FNC efforts to deliver agricultural services as widely as possible, only 42 per cent of conventional growers receive training in any given year, compared to the much higher 72 per cent amongst Nespresso AAA counterparts and 91 per cent for those selling to Fairtrade. Regarding good agricultural practices, farmers collaborating with value added initiatives showed statistically significant higher levels of adoption, which were 80 per cent for Nespresso and 95 per cent for Fairtrade.

Yet, it remains to be assessed what the Nespresso AAA Program can do to boost ownership and increase the extent to which better agricultural protocols become commonplace amongst its farmers. A similar evaluation is also required for farmers selling their coffee to Nescafé®. Knowing which factors are motivating changes in farming practices may help to realise the benefits producers can potentially reap. This, in turn, may improve the overall standard of living.

As part of the CRECE evaluation, surveyed farmers were also asked to give their perceptions on ten different parameters and state whether these criteria had worsened, improved or remained unchanged. Overall, farmers part of the Nespresso AAA and Fairtrade initiatives believe they have seen positive progress in their economic situation, income, farm management practices, trading opportunities, farm and village's environment, family health, employee and community relationships and quality of home life. For Nespresso AAA farmers, the most noticed benefits have been in terms of income, economic situation and trading opportunity. Farmers collaborating with Fairtrade indicated that the most perceptible developments had been experienced in farm and village environments, employee relationships, farm management practices, quality of home life and relations with the community (CRECE, 2011).

This information points to the strengths of the Nespresso AAA initiative, namely economic sustainability, and highlights the need to take a closer look at ways in which the programme can better address the various lagging social indicators. Interestingly enough, and despite being the parameters with the lowest scores, socially, both AAA and non-AAA farms have witnessed noticeable progress. In 2009, non-AAA producers obtained 46 points as opposed to Nespresso farms that scored 54. Two years later, the gains for control groups had gone up to 53 points and those attained by AAA farms reached 65. In 2009, there was a 17.4 per cent difference in the social index of AAA and non-AAA coffee farms. By 2011, this gap had widened to 22.6 per cent.

This means AAA farmers have formed more household assets, workers use more protective gear items and their living conditions are becoming more attractive. It is worth emphasising that the use of attire protecting farmers, and workers, from chemicals and other contaminating substances is an indication of the emphasis coffee producers are now awarding to safety and health. These individual behaviours and farm practices have considerably improved since Nespresso AAA has been raising awareness amongst coffee growers. Moreover, the perceptions of the household's quality of life and relationships with the employees have also improved during the past two years (CRECE, 2011, 2013).

The municipalities of Rosas and La Sierra exemplify the extent to which coffee processing areas collaborating with the Nespresso programmes have increased investment in coffee facilities and infrastructure. In 2008, these two coffee producing regions hosted a total of 1,930 Nespresso AAA farms: 1,009 in La Sierra and the remaining 921 in Rosas. That year, fermentation tanks were found in only 104 farms and solar drying systems available on a mere 77 estates. Only 73 farms in La Sierra had tanks and 39 invested in solar dryers. These figures were even lower in Rosas, where only 31 farms had fermentation tanks and solar drying systems could only be found in 38 properties. Three years later, these numbers had multiplied exponentially. In 2011, 469 farms in La Sierra were processing coffee employing fermentation tanks and 420 had solar drying systems. That same year in Rosas, 446 estates had tanks and almost an equal number, 424, had solar drying systems (FNC and Nespresso, 2013).

Similar achievements have been recorded in the 47 municipalities in Cauca and Nariño that are part of the AAA Program and where key initiatives have been put in place to boost sustainable and high quality coffee production. For instance, to February 2014, and in order to reduce water pollution, 1,998 wet milling wastewater ecological treatment systems and 5,232 household residual water treatment systems were installed. Also, 3,561 waste management systems were delivered and put into operation. In total, some 13,000 ha were renovated, more than 70 million plantlets delivered and 97 agronomist trained in farm sustainability, quality and profitability issues.

As an example of activities to improve producer competitiveness, 170 coffee farmers in Inza, Cauca, have received assistance to improve coffee processing infrastructure, support to make productivity-enhancing investments and training to improve agronomical practices. In Caldas and Antioquia a nursery was built to deliver 400,000 seedlings in 2013 alone. Through these renovation efforts, and to safeguard productivity and quality levels, rust-vulnerable Catimor coffee trees were replaced with rust-resistant Castillo® variety plants (Nestlé Nespresso SA, 2014f).

Nestlé's interventions in Nescafé® and Nespresso coffee procuring areas in Colombia have had encouraging economic and environmental impacts in the relevant communities. CRECE estimates that around 150,000 people in coffee growing communities have reaped social, economic and environmental benefits from the AAA Program alone. What is more, the initiative's positive impacts are not limited exclusively to advancing sustainability pillars. Nespresso AAA farmers have realised the largest benefits in terms of incomes, as well as economic and trading opportunities.

Average coffee growers and producers working with the Nescafé Plan seem to have more optimistic views of the future. Unlike their counterparts, many of the farmers selling their coffee beans to the Swiss company are moving on from mere survival mode to more sustained livelihoods, and also to higher profitability. This relates to the willingness farmers have expressed to continue their participation in the scheme and the demand that exists in other regions to extend the AAA initiative to more areas.

Technical assistance, in-kind help (for instance receiving rust-resistant plants free of charge) and the premiums farmers receive for good quality, verified and certified coffee have had positive impacts on their economic situation and on health and environmental conditions. Coffee growers are also able to visualise better the interplay of the different steps connecting their coffee plants, the practices they adopt to make them productive, their living conditions, the country's coffee culture and reputation, and the final product into which their work is transformed. AAA farmers are also better positioned to seize new market opportunities.

7 Lessons learned

Future challenges and opportunities

The demand for coffee and coffee-based drinks shows healthy rates of growth in Colombia and in international markets. In the long run, AAA coffee producers may be better served by looking at absolute production, and the overall quality of its beans, and not necessarily on relative market shares. In the International Coffee Organisation (ICO) list of countries by yearly coffee consumption per capita, Colombia ranks 55th, with only 1.8 kg. In comparison, Brazil has a per capita coffee consumption that is more than three times higher, at 5.8 kg per annum, which positions it at place 13 (ICO, 2014). Finland is the number one coffee drinker country in the world, with a per capita consumption of more than 12 kg every year. The next nine top countries by coffee consumption per capita are: 2) Norway (9.9 kg), 3) Iceland (9 kg), 4) Denmark (8.7 kg), 5) The Netherlands (8.4 kg), 6) Sweden (8.2 kg), 7) Switzerland (7.4 kg), 8) Belgium 6.8 kg), 9) Canada (6.5 kg) and 10) Bosnia and Herzegovina (6.2 kg).

It is to be noted that coffee consumption in producing countries is significantly lower than in importing ones, Brazil being the one notable exception. Higher domestic demand for coffee products in that country has been attributed to increases in disposable income and the retention of better quality coffee for internal consumption. Colombia has already begun to offer higher quality and premium speciality products to national consumers via gourmet brands and the proliferation of Juan Valdez® coffee stores and boutiques. In addition, boosting domestic consumption in Colombia can be achieved by stressing the links between coffee production and consumption. During the field visit carried out for this study, it was noted farmers and coffee drinkers alike frequently treated these two activities as wholly unrelated from one another. Instead, both coffee growers and coffee drinkers should be able to draw direct links between the coffee landscape and culture, which is deeply engrained in the country's narrative and fabric, and the final product they consume.

Furthermore, much progress can still be achieved to simultaneously capture bigger and/or more profitable business shares in the volumetric, specialty and ready-to-drink coffee markets. According to Euromonitor, demand for ready-to-drink coffee beverages has grown 8 per cent from 2011 to 2014, a market opportunity Nescafé® has seized by launching Shakissimo in Europe – a new range of chilled drinks (Nestle SA, 2014c). Similarly, regional coffees from specific geographical origins are emerging as yet another way to create differentiated products and as an additional opportunity to enhance the income of farmers. This is a growing niche market and opportunity for differentiation in the premium segment Nespresso is already looking into.

In the spring of 2014, the brand launched two limited edition varieties, the 'Colombian Terroirs', with coffee sourced from Cauca and Santander. Borrowed from the French wine industry, the concept of 'terroir' suggests that a region's temperature, sunlight and soil conditions, rainfall patterns, climate and topologies influence cultivation methods, coffee bean ripening rhythms and ultimately coffee's look, aroma and taste (Nestlé Nespresso SA, 2014b; Haynes-Peterson, 2014).

The Cauca Grand Cru comes from Colombia's south-western mountainous regions, where vegetation is lush, the weather equatorial and coffee is grown on slopes. Farms in this region tend to be small and thus to produce this limited edition, coffee was sourced from 14,660 AAA farms that on average were 2.55 ha. At the opposite end of the country, in the north-eastern dry, steep and mountainous region of Santander, coffee is grown under the shade of tall trees such as Guamo, Guayacan and Tachuealos. For this limited edition, coffee was sourced from some 1,389 farms that on average were 10.5 ha, almost twice as large than in Cauca (Nestlé Nespresso SA, 2014b). The production of these two special varieties alone brought economic benefits to some 16,000 farmers and is helping consolidate the brand's clusters in Cauca and Santander. There are some 15,000 farmers collaborating with the Nespresso AAA Program in Cauca alone. Their small farms represent some 16,705 ha of coffee plantations (FNC, 2013d).

Colombia's coffee sector, and its farmers, will benefit from FNC–Nespresso style collaborations to the extent that the brand's supplies originate from schemes like the AAA Program (FAO, 2013). Besides, Nestlé has the capacity to mobilise the knowledge, experience and best practices it has gathered in its 75-year long involvement in coffee production and its more than 125 years in rural development in different parts of the world. Policy learning options from this vast knowledge base are far from being exhausted. The company could also look all across the coffee sector, and in other sourcing countries, for successful examples of rural development from all around the world, which could be effectively used in Colombia after necessary modifications for specific local conditions.

According to the FNC (2014a), the AAA Sustainable Quality™ Program has become one of the highest impact, most successful and most comprehensive sustainability projects in Colombia's coffee sector. Despite such progress, structural and long-lasting improvements to producers' welfare require public action. In isolation, and regardless of its influence or economic power, no one player can or should take on state responsibilities, especially when that involves reforming an important part of the country's productive base and addressing rapidly changing social and demographic dynamics and the people's expectations. In fact, coffee production will only be profitable if it is acknowledged as an important part of the country's rural development strategy. Yet, many positive social and developmental expectations are attached to the Nestlé–FNC partnership, which is currently adding value to at least 13 per cent of the Colombian coffee harvest.

Local partnerships that work

Nestlé has established a long-standing and robust partnership with the most important institutional player, namely the Colombian Coffee Growers Federation (FNC). Since its establishment, this association has sought to organise and represent coffee producers, promote the efficient production of consistent good quality coffee beans, support the internal and external coffee market, promote social development in coffee growing areas, and develop Colombia's national coffee public policy (Cárdenas and Junguito, 2009). As probably the single most important coffee institution in the country, the FNC has built since June 1927 a common playing field for coffee producers and their families.

The Nestlé–FNC collaboration has helped to harness the sturdy institutional architecture and the knowledge base that the FNC has built over decades in Colombia's coffee growing areas. The Federation is making a determined attempt to spread and replicate good practices in different coffee growing municipalities. It provides the national, regional and local institutional architecture, social capital and trust needed to reach the farmers across the country. It is also developing or strengthening the local organisations required to implement the programme efficiently, providing inputs and commercialising the coffee beans, including assisting in the key process of traceability. Nestlé, for its part, provides technical assistance through the FNC, channels financial resources to execute programmes directed at improving the sector's productive infrastructure and environmental standards, and pays premiums to those growers supplying the finest produce (FAO, 2013).

This partnership has also contributed to accelerate the speed at which knowledge is transmitted to the farmers. It increases the sector's capacity to respond to large-scale challenges by pushing for local solutions and mobilising key resources in areas such as plantlets, inputs, greenhouses and, most importantly, agricultural extension services. Nestlé has followed, and financially supported, the technical assistance programme the Federation has established for the coffee sector. In fact, the FNC has a vast and respected agricultural extension service that has been built up over decades and trusted by the farmers. Its agricultural support services team now has some 1,500 members.

In 2013, Nespresso alone financed 140 agronomists in Colombia who visited and trained farmers in quality, sustainability and productivity – three key elements that can substantially improve farmers' incomes and living standards (Nestlé Nespresso SA, 2011a; Nestlé SA, 2014b). More intensive and personalised technical assistance has also encouraged coffee growers to take new risks and change long-held coffee production practices.

The TechnoServe–Nespresso collaborative effort in Caldas is another example of the sort of partnerships that have been forged to better the life of small coffee producers by integrating them into global supply chains for high-quality specialty coffee. By joining efforts with farmers, the FNC, Colombia's main coffee exporter Expocafé®, a local co-operative and the Caldas Coffee Growers' Committee, TechnoServe and Nespresso have worked together to enhance farm productivity, develop new pricing and marketing mechanisms and improve overall business management. The success of this partnership resides in the common and holistic approach actors have built and that guides their actions through a shared strategy and towards a common vision.

From 2006 to 2008, 1,260 small farmers whose plantations added up to 10,000 ha were effectively integrated into Nespresso's supply chain. They were trained in best agricultural and production practices, assisted in upgrading and purchasing milling infrastructure, and trained in improving operations, key processes to raising and homogenising coffee quality. Price incentives were introduced to reward quality improvements and encourage farmers to change farming practices. Traceability systems were designed and implemented to improve transparency along the supply chain, reduce payment delays, transfer higher price shares directly to the farmers and build producer loyalty. When the AAA initiative was first introduced to the Caldas cluster, 60 per cent of an average farmer's production fulfilled Nespresso's quality requirements. Two years later, in 2008, farmers sold 95 per cent of their yields to the AAA Program at a premium price 18 per cent above prices for regular green coffee.

TechnoServe calculated that from 2006 to 2008, revenue per hectare of AAA coffee went from USD 2,705 to USD 4,284. Consequently, farmers' net income per hectare rose by 20 per cent as it went from USD 1,981 to USD 2,375. At the time of this study, the Caldas cluster comprised 21 per cent of all AAA coffee producers in Colombia. As a result of better coffee production systems, the brand made its procurement processes in the Caldas cluster more efficient and reliable, its supply base larger and more secure, the volumes for coffee supplied higher and quality levels standard (TechnoServe, 2009).

These examples of long-lasting relationships can be leveraged and their positive impacts replicated and scaled up. They also highlight some of the main lessons learned and achievements joint collaborative efforts can bring about to coffee growing communities. Public–private and social partnerships have proven key for the success of the Nescafé Plan. Better and more lasting results can be achieved if the engaged institutions and organisations are embedded in the communities of intervention and these are able to build and maintain trust amongst participating coffee growers. More crucially still is to acknowledge, identify and treat coffee growers and their communities as the most important stakeholder in any initiative.

To start with, it is crucial all partners realise, acknowledge and take actions that see farmers as the fundamental stakeholders in all interventions. Establishing an open and participatory dialogue with farmers is pivotal to avoid productivity and sustainability enhancing initiatives to be perceived as external impositions.

Communication avenues are critical to reach the quality, productivity and sustainability goals set by the AAA Program. Gaining farmers' trust can be otherwise very difficult, which could compromise the success of any new initiative. Higher levels of trust are pivotal to sustain productivity and quality gains as they translate into higher producer loyalty to the AAA Program. Higher quality translates into lower rejection rates and thus more coffee sold at premium prices. This creates a sense of belonging amongst farmers as they regularly see results and can be seen as proxies for AAA Program ownership. They also reflect the coffee growers' willingness and commitment to carrying out best agricultural, management, labour-related and environmental practices.

For example, the production of the Naora Grand Cru was the result of a significant shift in individual and collective practices. Its success can be greatly attributed to the active involvement of coffee farmers, their willingness to take risks and change long-held practices as a result of the appropriate technical assistance services, engaging the right mix of partners, forging flexible and locally supported partnerships and working with coffee growers in collaborative, non-invasive ways. This Nespresso coffee variety exemplifies farmers' increased sense of ownership; the importance of monitoring how field operations are being perceived; the primordial role innovation and change play in improving coffee quality and raising its profitability; and the creativity, flexibility and diligence required to successfully ignite change and get coffee growers to support new practices.

Changing generational paradigms

The production of Nespresso's Naora Grand Cru is probably one of the most complex examples of the intricate relationships that have been forged between Colombian farmers, Nespresso consumers, Nespresso as a brand, Nestlé as a company, the FNC as a strategic and important ally linking farmers and coffee buyers and the research to test new coffee varieties carried out by Cenicafé. Launched in 2012, producing this limited edition variety required 1,600 farmers from Santander and Tolima to engage in a considerable exercise of trust in Nespresso and the technical advisors supporting its production. It borrowed from production, harvesting, processing and tasting methods more frequently used in wine-making and viniculture. Moreover, the idea, which took two and a half years to be executed, emerged from the innovative need to find new flavours and the creative experimentation with the new Castillo® coffee tree variety (Nestlé SA, 2012c).

As early as 1968, Cenicafé engaged in a vast research and experimental programme to genetically improve coffee plants, make them resistant to pests and diseases, and improve quality and productivity. As a result of such efforts, the Castillo® coffee variety was introduced in Colombia in 2005 on a large-scale (Cenicafé, 2011). It presents the physical characteristics of the Caturra variety, which is short with a thick core and plentiful secondary branches, and the rust-resistant strands found in the Timor Hybrid. The resulting type is high quality, rust and disease resistant; its cherries ripening without falling from the branches; and is composed of up to 80 per cent of larger-sized *supremo* beans (Coffee Research Institute, 2014; Alvarado-Alvarado et al., 2005). Within Colombia's grading system, in which coffee quality is associated with bean size and density, *supremo* beans occupy the top tier as they pass through the Grade 18 (18/64″ diameter) sieve. A higher share of large beans thereby benefits producers as they gain access to international markets and obtain prices paid for higher-grade, export coffee (Zecuppa Coffee, n.d.).

The production of Naora seized Castillo's® particular characteristics and implemented agricultural innovations that broke many of the coffee production paradigms farmers have followed for generations. Beans were harvested late, their sugar and acidity levels measured using wine-making techniques and their maturation prolonged for 15 days after the usual harvest timing. This meant cherries were over-ripened when they were removed from the branches (Cenicafé, 2012). Refractometers, instruments commonly utilised in viniculture for determining ripeness, were employed to resolve the moment at which the harvest was ready to be hand picked. Cherries were individually detached from coffee trees once they had reached a dark purplish red tonality, a stage where the sugar content had increased and cherries had developed notes of blackcurrant and blueberry, giving a unique and distinct coffee profile to Naora (Nestlé SA, 2012c; Nestlé SA, 2014b).

Producing this unique blend broke deeply engrained paradigms. Coffee producers had to trust the extension officers' expertise in this experiment. They, nevertheless, expressed having felt apprehensive throughout the agricultural process to obey new harvesting timings that contradicted their long-held knowledge about coffee production (interviews with coffee farmers, Colombia, June, 2013). For a normal harvest, waiting so long to pick the cherries would have resulted in coffee beans falling to the ground and acquiring an earthy taste (what is considered as a defect). It was also crucial to avoid risks of fermentation or mould. Farmers were thus facing new and higher risks that could jeopardise their livelihoods. As such, coffee growers engaged with this scheme received an extra coffee quality premium.

This commitment to produce a long-matured Grand Cru also led to the establishment of a long-term partnership with Nespresso and a more permanent brand presence in the region (Nestlé SA, 2012c, 2014b). Nespresso has since then expanded its portfolio in Colombia by including Santander as the sixth coffee cluster in the country (Nestlé SA, 2014b; FNC, 2014a). The programme has expanded its geographical reach, scope and number of participants in Colombia, and other countries as well. Recording, systematising and understanding the potential such additional benefits have to reshape rural livelihoods, economies, landscapes and living conditions will thus become an imperative in the near future.

Santander offers an important comparative advantage in terms of environmental sustainability and livelihoods resilience since 78 per cent of the coffee growing area in this department is planted with rust-resistant Colombia and Castillo® varieties. Santander has the country's highest percentage of coffee producing area with rust-resistant varieties and consequently the lowest rust incidence, which is less than 1 per cent. This obviates the use of agrochemicals to control the plague and makes farmers there less vulnerable to suffering productivity losses caused by this fungus (Café de Colombia, 2012).

The production of Naora called for thorough support from FNC–Nespresso extension officers to institute new agricultural processes, shift production paradigms and change handling procedures. Igniting change of that magnitude, and in such short time also, put to the test the partnership the Federation and Nespresso have for many years worked to establish. The proposed methodology made abrupt and counterintuitive changes to well-established and long-held harvesting practices. Information had to be effectively and timely transmitted from Nespresso coffee experts to FNC's officers to coffee plantation owners to farmers to pickers. Inter-institutional communication became determinant and imperative to persuade reluctant and sceptical coffee growers to undertake the challenge.

Agronomists had to engage in effective conversations with individual farmers, seasonal workers, co-operatives and local coffee committees to convince them to take part in this project and to welcome risky, yet profitable new practices (Nestlé SA, 2014b). The work of extension workers became even more crucial, their support responsibilities greater, and the pivotal role they play in assuring adequate programme implementation even more essential. The success of this endeavour once again highlighted the primordial place technical support occupies in all strategies aiming at veritably promoting rural development, advancing sustainability, and making coffee production an innovative, dignified and profitable occupation.

At the core of competitiveness: relevant technical assistance

By working with the FNC as a pivotal partner, Nestlé – through Nescafé® and Nespresso – has become an important part of the solution to the difficulties and challenges faced by Colombia's coffee sector, both at present and in the future. In return, the FNC has received financial, physical and human support to implement actions advancing its long-term commitment to improving the lives of the Colombian coffee growers. Once known for its professional competence and ability to understand and act in the international market, 11 years after the end of the quota system, the ICO warned of the Federation's inappropriate response to the challenges posed by the Colombian coffee sector's inclusion into the free market economy (Thorp, 2000).

At the micro and meso level, FNC's historical collective activities have given it a great deal of legitimacy, convocatory power, authority and avenues to channel community participation, as well as the institutional capital and presence to improve the quality of state investment. Unquestionably, the Federation has a proud past but its future role is increasingly uncertain given the rapidly changing coffee scenarios in the world and in Colombia itself.

Although it is true that a partnership with the FNC does bring credibility and trust among beneficiaries, it may also impede innovation, especially as this institution dominates the design and delivery of technical assistance efforts. Its size, number of members it represents, the vested and conflicting interests it looks after and its trajectory pose considerable challenges to instituting change.

Alarmingly, the Federation has not seen the urgent need to change the content and operations of many of the services it offers. Despite its long-term commitment to the coffee sector and its producers, or maybe because this role has been taken for granted, the FNC has not put in place the monitoring and evaluation procedures that it needs to maintain and escalate long-lasting and positive results. It has filled in crucial institutional gaps and has assumed many of the obligations of the local governments, especially at times of intensified conflict or as a result of weak governance. However, as Colombia's state has gained strength, presence and credibility, the paternalistic role local coffee growers' committees have played is undermining the responsibility and accountability of local authorities and hampering the development of local institutions in coffee clusters.

This old mindset clouds, and seriously constraints, seizing the many opportunities and overcoming the challenges that lie ahead for the Colombian coffee sector as a whole, and in particular for big players like Nestlé. The current private–public sector co-operation platform between Nestlé and the FNC has not thus far resulted in a forward-looking strategic plan for the Colombian coffee sector. Neither has it resulted in the sort of capacity building schemes that can make a difference in the lives and farms of those producers supplying coffee for the specialty as well as the mainstream market.

Historically accustomed to receiving FNC support, some interviewed farmers expressed their disappointment at how extension services have evolved. Once known for thoroughly aiding farmers, the relevance and intensity of support has diluted considerably. Interviewed farmers stated their disappointment at the nature of FNC visits to their plantations. The focus of these once looked forward to meetings has largely shifted to loan performance and repayment follow-up. Reversing this trend is essential. Technical assistance remains the most important mechanism to deliver to farmers the information, technologies and resources that will lead to improvements in production behaviour.

On many estates, farmers have gone back to artisanal coffee production. This leap backwards has been due to the interplay of numerous exogenous factors. Some of them include deteriorated technical assistance, farm fragmentation, rural labour shortages, changes in cultivation patterns, the introduction of cash crops, vulnerability to pests and diseases, limited access to key inputs and credit, low literacy rates, risk aversion, mistrust of new initiatives, reluctance to adopt novel methods and the advanced age of farmers. The complexity of the current coffee market is such that farmers no longer can depend on their own experience to plan and manage production. For coffee growers not effectively connected to local and global markets and information, overcoming productivity constraints is an even bigger challenge.

In this scenario, expectations of the content and subjects to be addressed during technical assistance interventions have risen. The effectiveness and quality of such support services is vastly dependent on the actual content of the material. The actors and institutions transmitting information on improved agricultural and farm management practices is an equally important factor determining how steep the learning and implementation curve is. Retention is vital. Unless transmitted knowledge is integrated into established systems and better practices institutionalised, any development capacity effort will continue to yield only partial gains. Boosting agricultural production also requires engaging with stable partners, buyers and capacity builders. Their presence sends a strong signal to coffee growers that their sustained efforts, diligence and investments (capital, social and physical) will offset potential risks and will ultimately pay off.

Farmers are used to receiving support but they demand such assistance adapts to their needs but does not oblige them to sell their coffee to those delivering agricultural services. This latter issue is one of the comparative advantages the AAA Program offers as it is proof of the brand's long-lasting commitment to the region, its farmers and the initiative. Unlike other schemes, the technical assistance offered by the Nescafé Plan and the AAA platform is not conditional on programme participation. This landscape approach to cluster building and quality improvement greatly departs from traditional initiatives that focus on delivering agricultural support strictly to farmers selling coffee to them. Instead, agricultural support seeks to raise the quality floor for an entire region, significantly enlarging the area from where coffee fulfilling AAA quality requirements can be obtained.

Although enthusiastically welcomed by coffee growers in the areas of intervention and demanded by farmers in other regions, delivering commitment-free, agricultural and farm management advice delays and weakens loyalty-building processes. Despite igniting positive changes in a community as a whole, rather than singling out individual farmers, this technical assistance model urges the brand to find alternative mechanisms to build and strengthen loyalty with supplying farmers. One way in which farmer to brand linkages can be built and strengthened is by seeking the support of long-established and trusted organisations, such as co-operatives. These organisations also consider programme continuity and trustworthiness to be indispensable to motivate farmers to institute changes in farm management. Since co-operatives are firmly engrained in the coffee communities they and their members belong to, they are excellent avenues to encourage farmers to implement and put into practice the suggestions, new techniques and novel practices suggested by AAA extension workers.

AAA and Nescafé Plan-delivered technical assistance should send farmers a definitive message that they are not being left alone to their own devices, especially given the landscape approach to coffee production these platforms follow. Especially AAA farmers know compliance with sustainability parameters is an incremental process the brand will help them with, yet they want constant reassurance that the AAA Program will exist for a long time, if not permanently. Doubts amongst coffee growers collaborating with the Nescafé Plan about the continuity of technical assistance services and the relevance of their applicability have remained despite public commitments to support farmers in Colombia to 2020 in its first stage. Both Nescafé® and Nespresso could communicate to their supplying farmers that their initiatives are still in their initial stages, to be followed by subsequent schemes in the future. Besides wanting higher and more stable base prices and premiums for their crop, farmers are looking for long-term partners, sustained investments and greater market access for their crops.

The 140 agronomists collaborating with the AAA Program in 2013 alone have different profiles that set them apart from their counterparts working with other initiatives or with the FNC. The portfolio of services they offer, and in which they have to be proficient, has to include farm management, profitability and environmental sustainability-oriented issues. They have stressed the need to make technical assistance content and delivery mechanisms the result of accumulated knowledge gains. This was identified as a possible way to design and implement consistent strategies through the coffee growing and processing cycle. Consequently, the technical assistance Nespresso facilitates substantially departs from the services ordinarily delivered by the FNC in one crucial way: it establishes a direct link between agricultural knowledge, better produce quality, enhanced productivity, sustainable practices and access to new and more profitable markets.

Additionally, Nespresso agronomists can offer more intensive, more thorough and more frequent services than those offered by extension workers collaborating with the FNC. Nespresso technical advisors follow up an average of 280 farms each year as opposed to a FNC technical advisor, who is in charge of at least 520 farms. More agronomists means more coverage, more thorough assessment of current practices and areas of improvement, more personalised assistance to make necessary changes and more detailed follow-up to assure farms are on the right track to boosting productivity, raising quality and advancing environmental sustainability aims. These gains can also be accelerated, better recorded and monitored and scaled-up due to the technological devices the AAA Program has incorporated as part of its activities. Technical advisors are using electronic devices (iPads and tablets) to gather information, furnish the AAA database, give feedback to both farmers and the brand as well as to speed up the analysis process in order to make the decision making process more relevant, timely and efficient. Compared to the FNC, the AAA scheme is still a young initiative but it has already proven it can ignite important changes in Colombia's coffee sector.

For decades, the FNC has supported farmers to improve farm management. Unfortunately, efforts to change coffee production practices have not paid off as coffee growers have not gained access to new or more profitable markets. Many farmers are not even aware of what happens to their crop once it is sold as green coffee. In contrast, the AAA Program has directed its technical assistance schemes at teaching coffee producers how their farms are the departing point in Nespresso's supply chain. To do this, even before aiming at changing farm management practices, coffee growers gain the skills and knowledge they need to be able to determine whether producing coffee is a gainful endeavour for them to undertake. When farmers obtain access to Nespresso as a potential coffee buyer, they know the improvements in terms of sustainability, quality and productivity the brand requires from them are targeted at gaining access to a final market. This has allowed farmers to draw a direct link between coffee production and its consumption, oftentimes in countries opposite to the world from where the beans were harvested.

Unlike other initiatives that take a checklist approach to verification, certification and sustainability, the AAA platform consists of constantly fulfilling and bettering their performance against a series of parameters, with the support of Nespresso technical advisors. As such, farmers not only acquire mechanisms to optimise the use of inputs, capital and resources. Guided and assisted by the applied Tool for the Assessment of Sustainable Quality, TASQ™, discussed above, farmers gain basic, yet thorough, knowledge about their own farms, how they are managing them, the areas where quick gains can be reaped and the processes in which they require the most support. Coffee growers have also become greatly aware of the importance attached to producing high quality coffee, the role they play in the supply chain and how the final product looks. They now know higher quality coffee obtains pecuniary recognition, as it sells for higher prices, and it is also awarded higher consumer value, as they are willing to pay more for it.

During the team's field visit to Colombia's coffee growing belt and to farms supplying their coffee to various initiatives, it became evident that the management of farms collaborating with Nespresso, Nescafé® and mainstream buyers is substantially different. Nevertheless, farmers have not always been able to tell how different technical assistance programmes differ from one another. For instance, even when the Nescafé Plan strives to deliver sustainability tools to coffee growers, for many of the farmers this research team spoke to, it was not entirely clear how the Nescafé Plan-facilitated technical assistance is different, in content and delivery mechanisms, to the services they have traditionally received as FNC members. The support the Federation has procured to more than 563,000 coffee producing families has for decades focused almost exclusively on traditional issues like general agronomic affairs. Social and business-related issues are still to receive the same level of attention as farm profitability and environmental conservation do.

Moreover, although it holds true that the quality and frequency of technical assistance services offered to farmers will greatly depend on the coffee market they sell their produce to, the knowledge gap between mainstream farmers, many of them supplying coffee to Nescafé®, and speciality coffee growers, collaborating with Nespresso, is widening considerably. The bulk of farmers, producing regular coffee for the mainstream segment, do not have access to the knowledge, nor safety nets to apply it, that can facilitate the lasting transformative changes they need to attract the younger generation to stay in the coffee business. The current FNC-driven technical assistance delivery process is deficient in imparting knowledge as to how best to improve the farmers' economic and living conditions, and how to make small- and medium-size farms transition from survival to profitability and sustainability. These are important issues on which the farmers need help urgently, but clearly this is a task a private corporation such as Nestlé cannot and should not be expected to undertake alone, especially when it comes to addressing deeply entrenched social dynamics.

Both AAA and Nescafé® support does include social parameters. Yet, these social programme elements refer mostly to labour-related practices. Little attention has been awarded to intra-farm and intra-household resource and income allocation, as well as to family members' gender, education level, participation in decision-making processes and recognition. Getting involved with family level dynamics is an evidently complex process but one that has pivotal implications to farm management and subsequently needs to be taken into consideration. For example, for most of the study team's field visit, women were rarely present at group discussions, their participation in collective conversations was timid and their role in coffee production unpaid and not explicitly nor publicly recognised.

In spite of the large share in productive work they do, women are very seldom included in decision-making processes, at the household, community or co-operative level. More rarely still are technical assistance schemes, rural development initiatives and agricultural policies targeted at them. It stands to reason that if women gain the proper skills, capacities, knowledge and visibility, not only should coffee productivity rise but also overall household living conditions improve. Women can be encouraged to join co-operatives and form women coffee growers' associations, be trained as extension workers, and gain even more elaborated skills such as coffee tasting.

The Nescafé Plan and Nespresso Program have already delivered important benefits to coffee communities in Colombia and thus gained the trust of people and institutions there. This formal and informal social capital could be harnessed to rally the necessary local support and start collaborative efforts to make way for interventions that tackle more complex social dynamics. Choosing the right partners to work with is equally crucial to addressing more structural and multifaceted challenges. Consequently, Nestlé needs to look for entrepreneurial, proactive, dynamic, forward-looking and more like-minded associates with whom to work in the coming years. At the very least Nestlé needs to nudge FNC out of its comfort zone and past glory, and help to develop a mindset which would address the important challenges the coffee growers are facing at present and likely to face in the future. The company would also benefit from capitalising on its partnership with the FNC in ways other than just hiring technical advisors.

Unfortunately, the existing institutional set-up that manages coffee production and marketing in Colombia has spent the last 20 years looking back at the golden coffee years during the quota era, and the high coffee prices at the time of the coffee boom. The FNC and the institutional architecture it heads (including the National Coffee Fund, the deposit warehouse company Almacafé, the exporting firm Expocafé® and the research centre Cenicafé), many of the farmers these organisations group and represent, and a myriad of private actors that are part of the coffee trade have spent valuable time and resources looking at the relative ground the sector has lost, and that its competitors have won, in the international market.

Nespresso's experience in particular exemplifies how quality, productivity and sustainability-oriented practices have brought about encouraging, replicable and positive impacts to individual farmers, coffee producing communities, the local environment and the brand itself. For these benefits to be permanently embedded into coffee production, technical assistance will have to be non-intrusive, high quality, timely, relevant, up to date and cost-effective. It is thereby pivotal that training and capacity building work towards embedding social, environmental and economic gains and improvements as part of coffee production and farm management. Veritably transformative and scalable change can only be the result of this long-term process of building and strengthening human capital.

The current arrangements and attitudes have yet to reflect adequately in terms of new and effective solutions for the challenges that the coffee farmers are facing today. FNC needs to reflect seriously on the adequacy and appropriateness of its current approaches, efforts and institutional capacities and managements to find effective avenues of devising and implementing the types of organisational, training, marketing and productivity-enhancing innovations that may be able to revitalise Colombia's coffee sector. It is imperative that local entities – public, private, academia and civil – develop the local capacities to support farmers to migrate to higher quality, higher value coffee beans.

New forms and areas of collaboration can help to formulate creative and resourceful ways of addressing the major structural challenges that Colombia's coffee sector now faces. The multi-layered nature of the challenges faced leaves ample room for all involved stakeholders to use their expertise and capitalise on their comparative advantages to pose previously unthought-of questions, devise out-of-the-box solutions, implement programmes and work synergistically together. During a conversation with CRECE researchers, the AAA Program was credited with taking the best of the organisations it works with and capitalising on the strengths of each player it partners up with. This factor is behind the interest coffee growing communities have expressed in partnering up with stakeholders and shareholders that can contribute towards their efforts to capture development aid, attract investment and resources. As an example, Nespresso's partner network has been a key factor of success as exemplified by the quality, environmental and income gains generated by collective schemes such as the Central Mill in Jardín.

The tenth anniversary of the AAA Sustainable Quality™ Program in 2013 has proven to be a timely occasion to forge new partnerships with the academic sector and organisations working to secure better economic trading and living conditions for farmers and workers. In June 2013, Nestlé Nespresso and Fairtrade Labelling Organisation announced a partnership to work together towards improving the lives of small coffee farmers in Colombia. FLO is a reputable organisation that promotes an alternative approach to conventional trade in which producers, traders, consumers and businesses forge partnerships to improve terms of trade and conditions for farmers. This partnership is interesting because it shows that once seemingly antagonistic groups are more willing to take risks and start working with each other.

Fairtrade's integral developmental objectives include granting market access for marginalised producers by shortening trade chains, establishing sustainable and equitable trading relationships, building capacities and empowering producers and raising consumer awareness and allowing them to be advocates for reforms to the international trading system (World Fair Trade Organization and Fairtrade Labelling Organisations International, 2009). Following this set of core principles, Fairtrade seeks to alter power asymmetries in trading relationships, market volatility and other structural conditions that have not helped farmers and producers to lead better lives. By working together, through the AAA Farmer Future Program, Nespresso and FLO aim to boost produce quality and farmers' quality of life. It also seeks to introduce welfare provisions for farmers and their families, including accident insurance, protection against price volatility and retirement planning (FLO, 2013; Nestlé SA, 2014b).

This welfare-oriented initiative, the AAA Farmer Future Program, will strengthen the brand's efforts to improve social conditions in coffee growing communities. At the same time, Fairtrade's core principles will be more widely shared (FLO, 2013). Together, Nespresso, the Colombian Ministry of Labour, the Coffee Growers' Cooperative in Aguadas in the region of Caldas, Cafexport, Expocafe® (in representation of AAA co-operatives in Caldas) and Fairtrade International have established a public–private partnership to pilot a retirement scheme for coffee farmers. A scant 2 per cent of them are currently affiliated to any sort of retirement or pension scheme.

In its initial phase, this set of diverse actors will provide a retirement fund to over 1,000 farmers taking part in the Nespresso AAA Sustainable Quality™ Program. The scheme will capitalise on the government-backed Periodical Economic Benefits national retirement fund (*Beneficios Económicos Periódicos,* BEPS) that allows Colombian workers to make flexible and voluntary contributions towards a retirement plan. It is hoped that this long-term initiative to provide social benefits to farmers will also encourage younger farmers to join and remain in the coffee business (Aguadas in Caldas, 2014; Nestlé SA, 2014a).

Public, private and social actors will be using their expertise to implement this pilot project. For instance, with this scheme, the government is expecting to bring a significant social transformation to Colombia's countryside and an ageing coffee sector. In this first stage, the Nespresso Farmer Future Program will mobilise a network of 33 agronomists to offer this retirement plan option to some 1,200 AAA farmers. If successful, the initiative will then be extended to the AAA farmers in the Department of Caldas, other regions supplying coffee to Nespresso and potentially to the more than 40,000 coffee farmers that are part of the AAA Program in Colombia (Ministerio del Trabajo, 2014).

Furthermore, Nespresso has made the commitment to source an increasing share of its coffee from Fairtrade certified co-operatives. This new AAA–Fairtrade alliance will provide small coffee farmers with greater income security and will work with co-operatives to implement social welfare projects for coffee growers and their families, such as accident coverage and retirement planning. Before this agreement was reached, the brand was already sourcing beans from five FLO certified co-operatives, grouping over 7,000 farmers in Caldas. With this new collaboration, the number of farmers that will be able to take part in the Nespresso AAA Program and will also receive the associative support of FLO to implement community-oriented initiatives will increase (FLO, 2013).

Nestlé, through its two well-renowned brands Nescafé® and Nespresso, is keen on tackling issues of generational sustainability in Colombia. Notwithstanding the positive impact the Farmers Future initiative may have, securing the next generation of coffee farmers requires structural changes, national and regional initiatives and collaborative efforts with traditional and external stakeholders. Above all, it necessitates innovation and piloting to devise initiatives that adapt to local realities, make the sector more profitable and offer greater economic alternatives to young people. Contributions to solving some of Colombia's challenges in the coffee sector have been generated locally and abroad.

In 2013, Nespresso started an annual global MBA Challenge together with its academic partners, the INCAE Business School in Costa Rica and the Centre for Sustainable Market Intelligence (CIMS). In total, 32 academic institutions from around the world joined the solution-formulating impetus through a MBA case study competition on Sustainable Colombian Coffee Farming (Nestlé Nespresso SA, 2012a). On one hand, this global initiative is likely to further strengthen the dialogue and linkages between the private sector and academics interested in sustainability issues. On the other, this international contest invites a new generation of business leaders to contribute to expand the brand's commitment to sustainability of the specialty coffee market (INCAE, 2013) and to formulate creative and innovative solutions to future supply chain challenges (CIMS, 2013).

The topic for this first edition of the prize called on contestants to design a framework aiming at 'Securing the long-term sustainable future of coffee supply in Colombia'. The winning team, four MBA students from the Rollins College's Crummer Graduate School of Business, Florida, proposed a 'Grow Forward' strategy to diversify farmers' income and mitigate seasonal employment (Nestlé Nespresso SA, 2013a). A more profitable and sustainable sector is likely to attract and retain the next generation of Colombian coffee growers. In September 2013, the team travelled to Colombia to assess the feasibility and potentially implement its multi-tiered approach to ensure farming is profitable and sustainable for at least the next 25 years.

A second edition of the Nespresso Sustainability MBA Challenge was launched in 2014 encouraging business schools around the world to come up with novel, creative and innovative ideas to further the Creating Shared Value Strategy. This concept, developed by Harvard professors Michael Porter and Mark Kramer in 2006, has guided Nestlé's efforts to simultaneously create and share value for the business, shareholders, factory workers, supplying farmers, service providers and consumers. Over 70 schools took part in this academic and intellectual exercise aiming at 'Pushing the boundaries of sustainable quality coffee sourcing' (Sustainability MBA Challenge, 2014).

These challenges are examples of the sort of new partnerships that may be able to invigorate the Colombian coffee sector. They also show that the new generations of business leaders are increasingly becoming more socially concerned and committed as employers, consumers and investors. Engaging younger people on and off the coffee farms may bring about fresh ideas and put forward the mindset of a new and hopefully sustainability-oriented generation. This may prove to be a way forward to tackle inter-generational challenges faced by the coffee growers not only in Colombia but also perhaps in many other coffee producing countries.

Capacity building

Nestlé is playing an important role in generating solutions to some of the most pressing issues confronting the coffee sector, and thus affecting one of its most important supply chains, not only in Colombia but also in other parts of the world. It may now be necessary to consider that this role may also include posing questions and framing problems in such a way that they can be effectively addressed. The State must play a leading role to ensure appropriate policy frameworks and provide enabling environments within which various private sector organisations, such as Nestlé and others, and the coffee growers can play important and meaningful parts.

The coffee sector can be strengthened and the lives of its farmers improved in more than one way. This is a powerful reason to increase shared value, as well as responsibilities of all the relevant stakeholders, and increasingly share value, with its producers, consumers and shareholders. The company should consider tackling challenges from as many points of view as its expertise and economics permit, in partnership with different stakeholders and looking ahead into different timeframes and different geographical areas.

Building farmers' capacities is a key area where Nestlé could consider funnelling its corporate strengths and expertise and take a more proactive and leading role in setting the technical assistance agenda in the communities it procures coffee from. After intensive field research, and an extensive desk review and series of analyses, the study team concluded that the content of capacity building programmes and their delivery processes have remained practically unchanged during the last few decades. This is in spite of the fact that the ways and channels to convey the messages have been adapted to new information and communication technologies. As such, there is an urgent need to restructure technical assistance to make the given information more specific, specialised, relevant and more easily absorbed and put into practice by the beneficiary farmers. At the same time, the delivery mechanisms have to be made more durable and the results more strategically oriented to avoid duplication and repetition.

In the view of the study team, Nestlé should review the benefits and costs of the present system of technical assistance that FNC is providing to the farmers and to what extent this is helping them to improve their financial and living conditions. Agronomists constitute the backbone of the Nescafé Plan, the AAA Program and Nestlé's comparative advantage in terms of raw material procurement. This is why, prima facie, it appears that Nestlé may wish to consider recruiting a new generation of extension agents who can provide the type of information and knowledge the farmers need in the future, rather than depending on FNC to deliver the services in the old-fashioned way as it has done over the past several decades. This would represent a radical change as to how agricultural extension services are delivered to the coffee growers in Colombia. Evidently, any major changes to the existing system would need to be carefully assessed, including their institutional and political ramifications.

Since 1866, and with activities spanning to 80 countries in 465 factories, over half of them in developing countries, Nestlé has amassed vast local and international knowledge and experience in the field of rural development (Nestlé SA, 2013a). In recent decades, the company has also accumulated vast knowledge and experience as to how to provide technical assistance most effectively and promptly to farmers in different countries of the world. In Colombia, and in many other countries, Nestlé factories have triggered the creation of quality-centred entrepreneurial clusters around their supply chains. In addition to this crucial know-how, the company has the financial, physical and human capital to innovate, create and consolidate its role as a leading player in the food and beverages sector in Colombia and in the world.

The company is also present in, and sourcing coffee from, countries that are Colombia's competitors in the coffee market and are coming up with innovative solutions to address geographical and production related challenges similar to those Colombia also encounters. In Brazil, output increases have been the result of better export and domestic prices paid to producers, higher productivity, readily available credit access, innovation and research, technological diffusion and improved quality. Brazil claims these productivity gains have been attained within the most rigorous labour and environmental legislation of any coffee producing country.

Despite the already high level of international competitiveness the Brazilian coffee sector exposes, its farmers are confident that they can still push the sector's productivity possibility frontier outwards. They are looking into investing in increasing yields in hilly areas, utilising slow release fertilisers and increasing mechanisation in mountainous regions (Brando, 2012). Colombian coffee growers and their associations, the Federation and its research arm

Cenicafé should all invest in developing, testing and scaling up similar alternatives. For its part, Nestlé could encourage and intensify information and best practice sharing amongst different country units.

Nestlé's pool of knowledge thus needs to be harnessed to its full potential to start pilot projects, compile lessons learned (good and bad) and scale-up successful initiatives throughout its supply chain. It could also be put into use to seize the areas of comparative advantages of other similarly proactive, forward-looking and flexible partners. Many of these good practices can be adopted specifically to suit the prevailing conditions of the Colombian small-scale coffee growers, or to update and upgrade the technical support farmers are currently receiving.

In the view of many interviewed farmers, the technical assistance delivered by FNC extension workers, some of them paid by Nestlé, is not very appropriate or relevant for the current conditions. Their realities have changed. New skills are required to face up to local challenges, national dynamics and global imperatives. In order that the farmers benefit in significant ways, extension workers need to focus on how to run small coffee farms as profitable businesses. As a first and pivotal step, this would make coffee production more attractive not only for the current farmers but also for those who will come after them.

Equally, the incentive system within which extension workers operate needs to be revised. The current system is based on activities, i.e., how many farmers are visited, how many workshops are conducted, etc. According to the FNC, in 2011, more than 1,800 people were working as part of the extension services and offering support and training in topics related to coffee production, productive plots, coffee growing areas and institutional arrangements. In the first nine months of 2011, the FNC had over 692,000 individual sessions with coffee growers, out of which at least 30 per cent were carried out on-site, directly on the farms. More than 28,000 group events were also organised and held, including meetings, tours, field days, demonstrations, courses, etc. However, these activity-oriented tasks give only somewhat limited information on to what extent the extension workers are currently making any difference to the incomes of the farmers, or quality of their lives (FNC, 2013b).

In contrast, a results-based management approach to agricultural support could become an important aspect of Nestlé's pillars to creating shared value that is currently conspicuously absent in Colombia. The average Colombian coffee farmer has at least three decades of experience growing this crop and yet has not been able to move from subsistence to economic resilience, let alone profitability. Nestlé thus needs to focus on upgrading farmers' on and off farm skills and, more importantly, building capacities on business management to transform coffee production. It ought to go from being the most profitable of the agricultural subsistence activities, to a genuinely sustainable and economically attractive rural development engine and catalyst.

Consequently, it is necessary to consider the success of the extension agents beyond the historical inputs and activities they carry out, irrespective of how important it is for them to visit and train the largest number of farmers possible within reasonable timeframes. Success parameters need to focus on outputs (the results of interventions), outcomes (short- and medium-term effect of those outputs) and impacts (actual changes in farmers' incomes and living conditions).

In fact, the Nespresso AAA initiative has already built business-oriented skills in other countries. Between 2010 and 2012, and in partnership with the local non-profit social investment fund Root Capital, a tailored financial and business management-training programme was delivered to six coffee co-operatives in Guatemala. Participating farmers received support and advice to make more effective financial decisions, obtain credit and improve the profitability of their farms. They were given access to financial management tools and gained knowledge about financial fundamentals (financial planning, reporting, controls, strategies and policies) and the development and management of internal credit systems.

These actions have helped coffee growers in Huehuetenango, Guatemala, run more profitable businesses, raise their income and improve the quality of life of their families. This initiative proved very successful and was then extended to Costa Rica in 2012. Similar knowledge gaps also exist in Colombia, where similar programmes could help to develop the capacities farmers need to turn their farms from subsistence plots to small, yet profitable, coffee businesses (Nestlé Nespresso SA, 2013e).

Quality gains and the benefits this brings to the farmers are still to be fully captured. The Nestlé group, through its two important coffee arms Nescafé® and Nespresso, needs suppliers who can reliably and consistently provide good quality coffee in large volumes. However, the sharp brand differentiation that exists between Nescafé® and Nespresso has led to the almost disarti-culated implementation of productivity-enhancing, quality-improving, social development-triggering policies and plans. Only 1 to 2 per cent of the world's coffee crops have the exceptionally high cup profile Nespresso requires for its Grands Crus. This poses an important sourcing constraint since even in AAA farms only around 30 per cent of each harvest meets those quality requirements (Nestlé Nespresso SA, 2011c). Nescafé® could purchase the good quality beans that do not meet Nespresso standards, instead of allowing a third party to acquire them. This would help the Nestlé group seize the gains of the many actions it has taken to improve and maintain quality and build a reliable, predictable and sustainable coffee supply chain.

Public, private and social actors supporting coffee production are rightly seeking to integrate farmers to the world of information and communication technologies. Cell phone and television penetration is already significant amongst younger farmers. Demand for Internet connectivity is growing steadily. Farmers are increasingly using online tools to make better-informed and more opportune commercial decisions.

In spite of the many benefits ICTs offer, technology alone cannot make up for undercapitalisation, underinvestment and knowledge gaps regarding agronomical and management practices. Issues requiring priority attention have not changed and neither has the technical assistance content and delivery mechanisms. Nestlé-funded and FNC agronomists and agricultural extension workers have to make front to some of the very same difficulties that have been affecting the sector for the last 25 to 30 years. Improving farm management – and the models followed to make plantations more effective, efficient and productive – has been a constant struggle. So has been creating a framework that accelerates and facilitates the adoption of productivity enhancing, economically attractive and environmentally sound agricultural practices.

Sharing the spotlight between farmers and consumers

Nescafé® and Nespresso have devoted extensive resources to its present and future consumers. These brands have developed products thinking of them and have quickly responded to their demands, needs, wishes, concerns and aspirations. For example, every year several new limited coffee editions are introduced. All of them have enjoyed substantial commercial success. Nespresso has also shaped consumer expectations surrounding sustainable coffee production.

Before coffee drinkers got increasingly preoccupied with the environmental and labour impact that result from its production, Nestlé sought to mainstream more environmentally friendly production practices and fairer labour standards throughout all its commercial operations in Colombia. In 2011, a few years after the first sustainability enhancing measures were adopted, the Nespresso Club Members expressed their interest in ensuring that coffee is sourced in environmentally and socially responsible ways in a customer satisfaction survey.[1] These so-called eco-committed customers like the good things in life but wish to enjoy them in a responsible way.

As a result, Nestlé has intensified communication efforts to let its customers, shareholders, Club Members and the general public know what actions the company is taking to make sure that each part of the supply chain is soundly formed. They are also being informed about how sourced raw materials, many of them procured many kilometres away from the final product, comply with corporately set ethical business practices. Nespresso has simultaneously gained more engaged Club Members and secured supplies of premium-grade coffee.

Naturally, these corporately endorsed and consumer-supported improvements in the company's business model have also benefited coffee producing communities. The environment is cleaner, natural resources are better managed and labour conditions have improved for hired workers. Nonetheless, the very same effort, resources and creativity that are allocated to trying to discern who the company's future consumers may be and what products they may need and want should be extended to encompass the coffee producers.

More attention needs to be placed on who will be the coffee growers of the future, how many of them will there be, how will their farms function and look like in the future and what sorts of skills and tools will they need to take advantage of constantly changing supply chains and consumer tastes and requirements. This is reflected in the findings of a study carried out by the

Sustainable Markets Intelligence Centre (CIMS), part of the INCAE Business School in Costa Rica. It was found that the AAA Program has certainly led to social and environmental improvements at farm level (CIMS, 2013).

Additional benefits are expected from the Real Farmer Income™ scheme and piloting initiatives to increase coffee growers' safety net, for instance through the Farmer Future Program. The brand's efforts to stimulate economic value creation have had encouraging outcomes but given the volatility of international markets and the many structural challenges faced by the sector, uncertainty for the coffee growers remains (Nestlé Nespresso SA, 2012a). There is thus a sector-wide sense of urgency to improve farmers' incomes and an even greater need to reshape relationships between stakeholders along the coffee value creation chain.

For the coffee sector in Colombia to thrive, farmers' net incomes to steadily increase for growers to lead dignified lives, and Nespresso to secure sufficient supplies of high quality coffee, farmers and brand will have to jointly address issues of on-farm sustainability and farm productivity. In 2008, Nespresso committed to focusing its activities to the Real Farmer Income™ approach that assesses farmers' business model and identifies ways in which coffee production can be transformed into a more profitable economic activity. Between 2008 and 2010, CIMS carried out an extensive study in Costa Rica, Guatemala, Colombia, Mexico and Brazil to determine the main factors driving farmers' income.

As the key factor determining farm income, productivity was identified to be in itself the result of the interplay of endogenous and exogenous variables, namely adequate farm management practices, the age of coffee trees, susceptibility to pests and the weather. Similarly important income determinants, prices and premiums were identified as secondary drivers of profitability and elements that are also closely tied to quality. In turn, it was established that the desired quality was attained by a series of factors related to productivity, processing and harvesting timings.

Notwithstanding the relevance productivity, quality, prices and premiums have for all farmers, no two farms are alike and thus no one single diagnosis can be made regarding the strengths, weaknesses and opportunities they face. Each farmer faces unique knowledge, human capital, capacities and practice gaps. There lies the biggest and most significant challenge for Nespresso, Nescafé® and any coffee player truly interested in and committed to improving farmers' wellbeing.

The actions Nestlé has undertaken in terms of economic, social and environmental sustainability has yielded positive impacts. They have, for instance, improved integration along the global coffee supply and value chain, both horizontally and vertically. Despite positive changes in farmers' living conditions and environments, scepticism prevails. Coffee producers, NGOs and civil groups worry that companies are only seeking to increase their control, albeit in new ways, over the supply chain. These fears are at least partially supported by the fact that the sustainability agenda seems to be driven by consumer countries and the multinational corporations headquartered there (Neilson and Pritchard, 2007). Coffee buyers – exporters, importers, roasters and consumers – decide what the parameters included in sustainability standards are. Producers, in the meantime, remain on the receiving end of key decision-making processes.

The potential benefits from enforcing private standards are clearly great. Yet, despite substantial work being invested into fine-tuning parameters and indicators, key stakeholders are not active partners. Farmers and their living standards are much talked about topics. Their role in decision-making processes, standard-setting and programme-implementation, on the other hand, is null. Sustainability-oriented schemes have created skewed governance structures that have failed to include the very same groups they seem to be advocating for. Nestlé and other NGOs and stakeholders setting private standards stress farmers are at the epicentre of these well-intentioned initiatives.

Farmers seldom know where they stand. They are told they are at the core of well-intentioned initiatives without even knowing it or being considered in their development and implementation. Their own countries uncritically import sustainability standards. Not one coffee growing nation has yet taken steps to manage or regulate such initiatives or to find mechanisms to harmonise global initiatives to national legislations and local realities. Not even Colombia, and its coffee institutional architecture, has adopted and adapted internationally set and accepted standards.

As an alternative to the plethora of certifications, seals, labels, verifications and standards stamped onto Colombian coffee, the FNC could look at the possibility of establishing collective national standards or guidelines worked out from 'below'. Guidelines applicable to all coffee produced in the country could also untangle the complex network of overlapping and inter-related standards. Colombia's traditional and praised institutionalisation could very well be harnessed to represent, organise and channel the needs of growers. This would ensure that standards work to address the social, economic and environmental gaps farmers themselves consider relevant. If substantial progress towards sustainability is to be attained, efforts should be at least partially driven by the interests of those producing coffee.

In Colombia, and elsewhere in the coffee producing world, farmers have little to say in the formulation, implementation and evaluation of the very same initiatives that seek to advance sustainability. Producers clearly benefit from partaking in such sustainability schemes. This book has compiled their testimonies and keen interest in carrying on working with the AAA Program and Nescafé Plan. Disappointedly, most of this participation has been passive. Insofar as these arrangements remain unaltered, coffee growers are excluded from processes where their interests are being sought after, and inputs from farmers are not included in the definition and implementation of sustainability standards, productivity and sustainability gains in the coffee sector will be transient and untenable.

Consequently, Nestlé, for both its Nespresso and Nescafé® brands, could work with local communities, local partners and local authorities to propose, plan and execute initiatives that take into consideration the needs of the host and the impacted communities and factor in the feedback of the so-called 'beneficiaries'. By including farmers in the standard-setting process, Nestlé's sustainability and creating shared value efforts would build the foundations for a more inclusive and equitable governance platform for coffee's supply and value chains. Such forum could connect farm-level realities with global supply chains, thereby establishing long-term sustainability goals. More importantly still, these interaction avenues would also make key strides towards building a shared vision of sustainable procurement amongst different stakeholders and shareholders.

For almost 70 years, Nestlé's industrial activities in five Colombian regions have certainly created value and brought many benefits to around 1,400 employees and about 7,500 suppliers. Looking into the company's future profitability and growth, value creation and distribution will have to be more evenly spread among the company's shareholders, particularly those who are at the base and beginning of the supply chain. Placing farmers at the core of any envisioned supply chain enhancing initiatives is one initial step towards achieving long-term company growth.

After all, for CSV to have truly transformational impacts on the company's value chains, all affected actors have to be actively engaged in defining the best available avenues for local potential to be turned into opportunities for business growth and social development. This process calls for the creation, expansion and strengthening of new avenues for regular communication and dialogue between suppliers, the public and private sectors and the society as a whole.

From activities to objectives and from transactions to interactions

The Nescafé Plan has focused a great deal on carrying out a series of activities and facilitating inputs that can promote rural development and increase the productivity of Colombia's coffee sector. From 2011 to July 2013, some 12.5 million high-yield, disease-resistant plantlets were distributed; 4,705 farmers trained and 6,800 people participated in 735 training sessions; and 5,146 farms visited to deliver personalised and specific advice on improved farming practices. These are substantial outcomes that have already contributed towards the achievement of certain outcomes, such as increased productivity and income or the reduction of water pollution. Nevertheless, for the Nescafé Plan to actually reach the goals it has articulated, it needs to focus on reaching outcomes and measuring impact. Especially at the field level, there is the risk of programme activities becoming ends in themselves.

Tracking, measuring and evaluating programme progress requires the same level of flexibility, innovation, comprehensiveness and responsiveness as planning and implementing the initiatives themselves. Indicators need to be reformulated and strengthened to reflect incremental changes. Taking a close look at the CRECE study, it seems that this demanding task to gauge impact is more easily performed when it comes to measuring economic and environmental results. Quantifying social impact and progress towards sustainability makes the measuring challenge even greater. Demonstrating and quantifying a connection or relation between an output and the advancement of a socially desired outcome is costly, time consuming and requires great data collection exercises throughout time. It is, however, a key process to gauge the success of an intervention and whether it is actually contributing towards improving farmers' lives and securing the sustainability of coffee production for many decades to come.

With time and fuelled by growing needs and demands for coffee products, the Nescafé Plan and Nespresso AAA Program have grown larger in geographies, scale, scope and complexity. In Colombia, and all the other countries where these schemes are in place, the number of collaborating actors and institutions with whom the company has established inter-organisational relationships has risen accordingly. Relational interactions have become, and are likely to remain, the most important mechanisms in building trust, confidence and continuity in the coordination mechanisms that join the different stages in the coffee supply chain.

Focusing on strengthening company–farmers interaction and rapport is another way to achieve lasting and transformational impacts as a result of which coffee producers can aspire to better living standards. This connection can be built and furthered by capitalising on the strong ties that already exist between technical advisors and the farmers they support. In other scenarios, the informal nature of many of these transactions and interactions could impinge on their longevity or robustness. This is one way in which Colombia is unique. Coffee institutions have kept very close contact with farmers through agronomists. They deliver new and enhanced agricultural, managerial and trade-related knowledge. These skills have the potential to go well beyond closing farmers' human capital gaps and making coffee production a more profitable economic activity.

Capacity building schemes rely on relational transactions and on keeping direct and close contact and rapport with individual coffee growers and the organisations they form. As a result, knowledge-exchanging processes can carve out and strengthen mutual and effective communication avenues between a coffee brand and its supplying coffee producers, coffee producers and technical advisors, technical advisors and buyers, and buyers and consumers. Establishing these types of long-term relational and social capital partnerships will greatly assist Nestlé, and its coffee brands Nespresso and Nescafé®, in securing the human, physical, social, technological and political capital the company requires to continue operating, growing and innovating within the coffee sector.

To a certain extent, this has already been achieved in Colombia. In the coffee-processing factory at Bugalagrande, Nestlé has established long-term social capital and loyalty-based, relational transactions with supplying farmers, factory workers, ancillary service providers and the surrounding community. It holds true that it is in Nestlé's interest to develop long-term relational interactions, and not only transactional ones, with the myriad of small-scale producers it works with. The stronger the trust, confidence and loyalty ties linking a company and its suppliers, consumers and the community they belong to, the more profitable its activities will be and its presence longer lasting. For farmers, benefits include partnering up with a reliable, reputable and renowned actor that can embed corporate growth prospects with extensive community development. To successfully match corporate objectives to local opportunities, endowments and needs, companies and suppliers have to work to build long-term, social capital based relations and move away from sheer transactions and merchandise exchanges. Building such ties is a challenge. Success requires strong stakeholder commitment and their continuous engagement.

The coffee sector has for long been characterised by transactional relationships established between different actors in the supply chain. This may provide a partial explanation as to why coffee growers do not appear to be very excited about the future presence of Starbucks in the country. In July 2014, the Seattle-based coffeehouse chain Starbucks opened its first shop in Bogotá. The company, which has been buying Colombian coffee for over 40 years, announced it will serve only locally grown beans in these new locations. This is a move the Colombian Coffee Growers Federation seems to be enthusiastic about, as the company has made public its goal to double to two million the number of coffee bags it acquires from Colombia every year. The commercialisation of one million extra bags of coffee beans will be paid at a premium up to 15 per cent above the current prices, which could potentially represent an income improvement for the beneficiary farmers.

Many farmers depend on the performance of differentiated coffee value chains to obtain higher incomes but a lot remains to be done to make these premiums reliable and predictable to make a real dent in their quality of life. More relevantly, farmers perceive the arrival of Starbucks as the substitution of one buyer for another, albeit a new one. Óscar Gutiérrez, one of the coordinators of the Movement for the Defence and Dignity of Colombia's Coffee Growers, is of the opinion that the business activities of this new player will not represent any material change in the coffee trade. Farmers are more concerned about the price of coffee and production inputs, problems with debt repayments and the extension of new credits, and the control of mining activities in the coffee growing regions (*El País*, 2013).

However, the presence of this new important global coffee player will undoubtedly intensify competition for coffee at the top tier of the quality pyramid. It is also very likely that the farmers already producing high quality coffee, and those aiming to sell their beans to Starbucks, will have to meet the terms of yet another company-specific certification programme. Farmer compliance with firm-specific standards, and the support the company offers farmers so they can fulfil them, creates mutually reinforcing relationships between buyer and producer. On the less positive side, it can also lock both parties in exclusive supply chains, raising the costs and limiting the opportunities of engaging with new partners, and reducing competition (Neilson, 2008).

Puzzling implications of human capital creation

In today's world, coffee growers have ceased to be only that. As part of value added global value chains, producers ought to become proactive and competent entrepreneurs. This would be a radical shift in farmers' profile that is likely to strengthen value chain governance for specialty coffees (Humphrey, 2006). Complying with schemes that may lead to an improvement of the welfare of farmers and their families involves acquiring, building and putting into use a new set of capabilities, skills and knowledge. These new dynamics may contribute to buyer–supplier relations that are significantly more comprehensive than ever before. Or, it could make the process more complex and difficult. Nestlé is giving farmers the tools they need to play this new role. Concurrently, competence gaps between mainstream and speciality coffee producers are widening, posing constraints to the future expansion, feasibility and rapid effectiveness of schemes like the Nescafé Plan.

The successful implementation of the Nescafé Plan will certainly help to increase productivity and total yields for 4C verified coffee. It remains to be seen if demand for specialty coffee in both producing and importing countries can be met solely by counting on the realisation of potential productivity gains. What is certain is that consumer demand will in turn intensify demand for beans fitting specific quality and taste profiles. More and better coffee will be demanded from hopefully more productive farms.

It is not to be overlooked that unless producers wanting to meet a certain set of standards get well-timed access to support mechanisms and technical assistance, incentives and access to capital and credit, competence gaps to comply with sustainable practices run the risk of becoming barriers to entry. This is especially relevant for smaller and mainstream coffee farmers, whose capabilities to partake in increasingly more competitive markets are lower and commercial channels to better their lives fewer. In this way, and added to challenges regarding labour shortages emerging from emigration and occupational changes, human capital differentials could compromise future coffee procurements. Lagging farmers are likely to take considerable time to move along the quality ladder and reach the level of sustainability required and expected to supply coffee to the Nespresso AAA Program.

This case study took a comprehensive look at how Nestlé, through the Nescafé Plan, has tackled challenges in the coffee supply chain for its Nescafé® and Nespresso brands. In Colombia, these schemes have assisted farmers in raising productivity and adopting more sustainable agricultural practices. Environmental, social and economic sustainability has improved as a result. Communities have benefited. Coffee producers are more knowledgeable about quality, productivity and sustainability. They are also better connected to global market opportunities. Whilst this is taking place in the field, the largest coffee buyer in the world is ensuring it is able to secure supplies for the highest quality coffee. Not only is Nestlé fulfilling its own internal commitments to source raw materials responsibly, it is guaranteeing a reliable inflow of good quality coffee beans to feed its expanding coffee business.

Note

1   Le Club concept was introduced in 1989 to create an upmarket image and give its members the notion of premium status. From that year on, members have been offered various services, distinctively the purchase of capsules from the Internet or at the Nespresso boutiques.

8   Further thoughts

Governance structures in the coffee sector determine the way and extent to which resources, gains, benefits and profits are spread along the value chain. In the prevailing system, economic profits are still concentrated in particular segments and a handful of firms have enough power to establish, shape and enforce the parameters that guide the operations of other actors who are also part of the chain.

Over the years, Nestlé Colombia seems to have used its quality-centred corporate comparative advantage and vast pool of successful experiences in rural development to address social needs, promote economic potential and contribute to environmental conservation efforts. This has shaped the way the company sees its suppliers. Instead of thinking of coffee growers as being at the bottom of the pyramid, they are seen and treated as an important base for the company's activities. This is helping the firm build interdependent, mutually reinforcing and reciprocally advantageous partnerships with myriads of smallholder coffee growers in Colombia. Farmers interviewed for this case study repeatedly expressed their desire to be included in, and remain part of, the company's productivity and sustainability-oriented schemes.

In this way, the company is forging the type of sustainable partnerships it depends on to reliably and consistently procure high-quality raw materials. Nestlé then uses these inputs to produce the coffee goods it is known for, as well as to keep itself at the forefront of innovation in the world's food and beverage industry. This model to deliberately build more competitive local contexts suggests risks are better anticipated and managed, operational costs reduced, the hazard of supply disruptions attenuated, efficiency enhanced and profitability heightened.

Participating farmers, for their part, are able to obtain higher and better yields; increase their incomes; become more resilient to unexpected and/or prolonged economic, social and household shocks; and ultimately improve their living standards. The aggregate effect of more prosperous, dignified and promising individual lives is a thriving, more competitive community; sustained growth for the local economy; and significantly enlarged opportunities for the new generations which would attract them to become coffee farmers. These are improvements all farmers would like to experience and that explain the popularity and demand for Nestlé's technical assistance programmes and support. Farmers want their coffee to reach the quality and sustainability thresholds that will make it eligible for obtaining quality premiums.

Strengthened societies can become sustainable development clusters with wider competitive frontiers and stronger, more flexible human capabilities and resources. These elements can be seized to build safety nets and pull forces against the type of economic, social, cultural and professional dynamics that are acting as migration push factors for many young people in the country. Assuring sustained, respectable and wilful generational relay will take much more than the initiatives of one private sector actor and the network of public, civil and academic partners it can rally. Unless more decisive measures are taken in concert with all the major public and private sector actors to support rural areas and the coffee farmers, production and procurement of coffee in the future is likely to be significantly compromised.

Farmers still require help in fully absorbing newly acquired capacities and putting them into practice on the farms. At present capacity building is rather repetitive and touches only a narrow range of issues despite the fact that coffee production is a vastly complex activity, and becoming even more complex with time. Most of the training sessions seem to revolve around the use of fertilisers and soil analyses that, albeit key to ensuring productivity and profitability, are not the only inputs or production aspects farmers should focus on.

Technical assistance could instead address the flat parts in the farmers' learning curve and effectively develop the capacities farmers need to formulate, manage, adjust and implement the sort of long-term business plans that can help them improve their economic and living conditions. Moreover, additional services could encompass the provision of long-term and opportune financing for farmers to purchase the capital goods and inputs as and when they need them.

An invigorated knowledge circuit is thereby crucial to transform coffee production processes in Colombia into a dynamic, market-responsive, productive and profitable activity. This long-term goal will be achieved by fostering generational relay, renovating old coffee trees, promoting the adoption of practices that improve profitability amongst coffee growers, support the productive restructure of marginal coffee growing areas, manage financial mechanisms that allow plantations to remain in good condition and incentivise the development of productive projects that can complement coffee growing activities. This will need a new approach to technical assistance since the extension workers of FNC with whom Nestlé is working at present do not currently have all the necessary knowledge and expertise to sustainably revitalise coffee production.

Skills will have to be developed amongst farmers and trainers alike, and targeted efforts will have to be undertaken to support capacity building and training activities that are oriented to the specific needs of women in the coffee growing business. As a whole, rural communities can greatly benefit from formal educational alternatives that can help address knowledge gaps, foster the involvement of younger people in farming and professionalise farming activities. It cannot be stressed enough that these are not activities Nestlé alone can implement nor can it be expected to do so. Government policies and actions are ultimately required for creating the necessary enabling frameworks to trigger structural change at scale even when social and for-profit enterprises have sizeable and valuable contributions to make to development.

What Nestlé can realistically and effectively do is to continue to mobilise and motivate its partners and keep on actively collaborating with relevant, competent and like-thinking actors. It can make use of its reputational, innovation and commercial muscles to encourage the implementation of public, private and social projects that foster the development of productive activities that generate value for rural dwellers and improve the coffee sector's physical and institutional infrastructure. Certification and verification schemes are examples of the existing partnerships and avenues available to farmers to make commercial, environmental, agricultural and social improvements to coffee production, add value to this agricultural activity and bring benefits to farming communities.

It can also do a great deal to encourage its consumers to be more actively engaged with the company's sustainability initiatives. Keeping close to its corporate mission, Nestlé can continue to mainstream sustainable practices throughout its supply chains as an answer and in anticipation to changes in consumer attitudes and concerns. This two-way approach to building and responding to consumer loyalty will contribute to attracting eco-committed consumers as well as to showing the average consumer that good quality products can be the result of social, environmental and economically responsible processes.

There is clearly still ample room for policy innovation to truly transform the socio-economic conditions of farmers and, with that, improve the standard of living of those whose livelihoods and future wellbeing are closely interlinked to the coffee beans. Some of the most important innovations have come from private and social actors in the form of private standards and sustainability initiatives. Certification and verification labels, compliance indicators and sustainability procedures like those promoted by Fairtrade, Rainforest Alliance, 4C and AAA have proliferated. For them to retain and build on their popularity, ethical muscle, reputation and awareness raising capacity, all of these schemes should be thoroughly evaluated.

In principle, these initiatives all share some core values, namely sustainability in coffee production and better livelihoods for coffee growers. It remains to be seen how this will be possible, especially as one fundamental element is still absent: farmers. They are yet to be actively included in how programmes are planned and implemented, what topics are covered, what are the most pressing concerns they, their families and their communities face. Most initiatives place farmers at the core of their actions but take little or no feedback from them. A shift in mentality is called for. Farmers could definitely use more targeted and effective technical assistance. But what they have stated they most pressingly require are partners who are willing to take them on board and open platforms and opportunities for them to shape their own future.

In our view, Nescafé® and Nespresso should consider building further on the successes they have already achieved on the Colombian coffee scene. Veritably comprehensive economic, social and environmental sustainable coffee production can bring about considerable progress at a scale that would receive the country's rural sector. This means, Nestlé, as the world's and Colombia's largest coffee buyer, should make use of its corporate competitive advantages, strengths in innovation and quality management, and emphasis on the development of relevant skills to reshape and revitalise the country's relationship with the world's most traded crop.

Future initiatives should focus on how best to ensure an economically and socially transformative coffee business in Colombia, especially those that would attract the younger generation to growing coffee. New farmers will need to find life in the fields an attractive, profitable and respected profession. This would truly create shared value for the company, its coffee growers, its shareholders and the millions of coffee drinkers that choose Colombian milds in Nescafé® jars and Nespresso capsules.

Economic profitability, environmental sustainability and social development cannot be fully achieved overnight. Nestlé's activities in Colombia for over a decade reflect this. Farmers, consumers, workers and shareholders alike are gradually, and constantly, reaping the gains from more reliable, better-quality coffee supply and value chains. Farmers in areas from where Nescafé® and Nespresso procure their coffee are harvesting some of the best crops amongst the already famous Colombian milds. The result so far has been encouraging. Economic, social and environmental conditions are better now than they were before the initiatives were rolled out.

The years 2013 and 2014 seem to have brought good news for Colombia. Global demand for coffee and coffee-based drinks is expected to continue its 3–4 per cent annual rise for some years to come (El Nuevo Siglo, 2014). The world's coffee prices are higher, the exchange rate in Colombia has become attractive for the farmers, global demand has intensified, rust leaf plagues are affecting Central American competitors, strong rains in Indonesia and Vietnam seem to be threatening growing conditions and road access and Brazil is facing a protracted drought. Significantly important for the world market, weather-related troubles in Brazil have triggered concerns about lower yields, leading to sharp, but not exorbitant, price hikes.

As an encouraging note, Colombian coffee production is already recovering. For the last nine years, the Nescafé Plan, in partnership with the FNC, has directed its efforts at increasing productivity and reaching historical output levels (Figure 8.1). By 2012, the coffee tree renovation programme had covered 500,000 ha with improved, rust-resistant, high-yielding coffee varieties. Two years later, aggregated national production numbers and the current state of coffee plantations in the country reflect FNC-led activities as well as support coming from the Nescafé Plan.

To 2013, 78 per cent of land under coffee cultivation had been renovated, amounting to some 754,000 ha. With rust-resistant varieties covering 58 per cent of the renovated areas, the incidence of leaf rust was brought down from 30 per cent to 4 per cent. As approximately 703 million new trees were planted, tree density went up by 8 per cent, from 4,642 trees per ha, on average, to 5,015 trees. In 2008, the average tree age was 12.4 years. Five years later, in 2013, average age had dropped to 8.2 years. As a result, in the span of just one year, from 2012 to 2013, productivity rose by 35 per cent (Muñoz, 2013).

By July 2013, coffee production was recorded to be at a ten year high of 1,031,000 bags, which represents a 54 per cent increase in coffee production from the levels reached 12 months earlier. From January to July 2013, the Colombian coffee harvest almost totalled 6 million bags, more than 1.6 million bags and 38 per cent more than the 4.3 million bags produced in the same period the year before. By 2014, output was 12 per cent higher. From July 2013 to July 2014, Colombia produced 11.5 million bags of Arabica coffee. This is a 27 per cent jump over the same period for the 12 months prior.

A mostly export-oriented sector, higher Colombian national production levels have naturally translated into higher volumes of coffee sent abroad. Coffee exports, estimated at 787,000 bags, were 42 per cent higher than the 552,000 bags harvested in July 2012. From January to July 2012, 3,954,000 bags were exported. In 2013, in the same period, exports were 30 per cent higher, at 5,138,000 bags (FNC, 2013d). Even when the full impacts of the productivity-enhancing efforts and tree renovation schemes are expected to be fully realised in the course of the next four to five years, these first signs of recovery in the Colombian coffee sector are a most welcome development for the 564,000 farmers of the country.

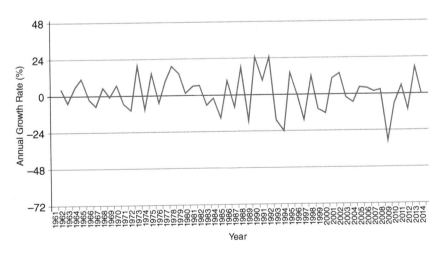

Figure 8.1 Colombia's green coffee production annual growth rate (1,000 60 kg bags)

Source: available at http://www.indexmundi.com/agriculture/?country=co&commodity=green-coffee&graph=production

Colombian coffee exports from October 2013 to the beginning of 2014 were 8.7 per cent lower than in the same period the year before (Anacafé, 2014). Even then, several years of surpluses and increased exports from Colombia, Uganda and India are cushioning production shortfalls in Brazil, the world's largest coffee producer (Wile, 2014). On 1 March 2014, Arabica was trading 54 per cent higher than its level at the beginning of the year whilst Robusta showed a more modest increase of 18 per cent (Ananthanarayanan, 2014), meaning the price gap between the two types of coffee has reached its widest in 22 years (Anacafé, 2014). Arabica coffee due for delivery in March reached USD 1.72, its highest point in 14 months on 21 February (Yang, 2014), which broke a price decline trend that had lasted since May 2011.

Although seemingly positive, the current coffee scenario is not unambiguously encouraging and no one can be sure how long it may last. Besides, this price surge reflects the almost inherent volatility in coffee production, agriculture's dependence on propitious climatic conditions, the role a limited number of countries play in setting prices and how world coffee supply can be met by established and emerging players alike. It remains to be seen what levels of stocks buyers, roasters and Brazil itself currently have, which can always affect world supply and thus prices. Colombia's stock levels are far too modest to affect the world market in any significant way. Because of the diversity of the suppliers, there are always players in the global market with the capacity to fill in any supply gaps. As such, the fundamentals of Colombia's coffee production should improve independently of temporary changes in the composition of world coffee markets. The country cannot, and should not, count on, or wait for, other producers' misfortunes to reform its domestic sector.

Public and private sector initiatives alike have already made considerable progress to address the many macro, meso and micro challenges the sector currently faces and is likely to encounter in the future. Although hopeful, gains are nowhere near offering widely spread and transformation solutions to the difficulties and barriers coffee producers encounter on a daily basis. Coffee price volatility, the ever increasing difficulties to predict climate patterns due to La Niña or El Niño phenomena as well as climate change, due to mounting production costs, declining productivity and production, labour shortages and higher prices for key inputs such as fertilisers and pesticides are some of the elements making farmers' lives rather uncertain and more precarious. These are all powerful reasons urging Nestlé, and all other interested parties in making coffee production in Colombia sustainable and productive, to move beyond the delivery of generic old-fashioned and primarily agricultural advice. The global supply and value chains farmers are part of make demands for a broader set of skills from them.

It has been previously mentioned that coffee production is progressively recovering, pushed by productivity increases. There is real potential for overall harvest figures to reach contemporary highs. Many of the recently renovated trees are still to reach productive age, improved agricultural practices are still to be fully entrenched in daily production practices and a few more years will have to pass before the spillovers of social programmes delivering public services to farmers are fully appreciated. There will be more trees in their productive age, each one producing higher yields. A natural question arises: who will hand pick all these cherries?

The second largest harvest of the year takes place between October and December, when some 12 million bags of green coffee are processed. Labour demands to collect at least 4,000 million kg of coffee cherries in those three months exceed 300,000 workers. Even before production levels began showing signs of recovery in 2013, labour shortages had already been felt for some years. In the upcoming years this cyclical problem is only likely to deepen. Although production increases result in intensified labour demand and higher wages, Carlos Armando Uribe, Technical Manager at the FNC, remarks that better remuneration may not be enough to attract sufficient workers for this and future years' harvests (Asoexport, 2014).

It was discussed earlier how agriculture is struggling to retain young people in rural areas. Rural youth is not only drawn into other better-paid economic and productive activities, such as construction and the informal sector. Despite high levels of urban poverty in Colombia, young people are pushed out of rural communities by the large gaps they perceive exist, and truly exist, between urban and rural areas in terms of public goods, education, health, social safety nets, infrastructure, economic opportunities and employment options. Antipoverty schemes will necessarily have to put the rural sector at the core.

The time to open up new niche markets for Colombia's coffee and deepen its commercial penetration seems right. Demand is escalating in the mainstream and specialty markets. It has been estimated that the global coffee retail value and volume sales for the 2012–2017 period will increase by 6.5 per cent in retail value and 2.5 per cent in retail volume (Flury, 2012). These are clearly good opportunities for the country, although not the only ones. The new coffee scenario, with more and different actors, more inclusive governance structures and higher international prices could constitute a turning point for Colombia. Seizing the opportunity at hand, the country's coffee institutions could channel their efforts and propose substantial solutions to the structural impediments that have pervasively compromised the quality and quantity of coffee it is able to position in the global market.

The country's new public policy on rural development is expected to tackle some of the underlying factors constraining growth in the sector. For a start, it could articulate institutional and governmental efforts to overcome the existence of islands of productivity and, instead, try to promote growth across the countryside. It cannot be stressed enough how important the participation of farmers is. They constitute the main share and stakeholder group in rural communities. Numerous as they are, small farmers lack the channels to effectively influence their economic landscape or shape the public policies that direct the courses of action, measures and priorities to be pursued in rural areas.

Closing development gaps between cities and rural communities is likely to be more effectively achieved if rural public expenditures move away from extending subsidies to specific producer groups and are redirected at enlarging the delivery of public goods. Subsidies tend to be regressive and inefficient. They also tend to see rural economies as detached from overall national welfare when, in reality, they are important development and growth engines. More importantly still, prosperity in the countryside is likely to accelerate the peace building process in Colombia. That long path will be less troubled, more solid, more inclusive and long-lasting if rural populations benefit from long overdue investments in rural education, infrastructure, health, social protection, research & development and targeted antipoverty schemes (De Ferranti, 2005; UNDP, 2011).

Productivity, quality, technical capacity, infrastructure development, labour retention, intra-generational relay, marketing skills and compliance with economic, social and environmental sustainability parameters are ever moving targets. Continuous improvement is imperative if the country wants to maintain and further its market position. If properly targeted, existing financial resources could be reallocated to tackle these pervasive challenges.

Starting from 24 October 2012, the government, through the FNC, began implementing a subsidy scheme to protect coffee earnings (Protección del Ingreso Cafetero, PIC). In just one year, 353,323 coffee farmers in 580 municipalities and 21 departments received more than COP 806,696 million (around USD 400 million) in support (FNC, 2013d). This year's prices paint a totally different picture. At least part of the COP 1 billion the government had allocated to paying production subsidies to coffee growers could instead be used to strengthen Colombia's coffee social, economic, environmental and institutional architecture and make it more competitive internationally in the future.

Even when competitiveness does use other countries as reference and comparison points, the Colombian coffee sector needs to prosper and offer better living conditions to the families growing the crop regardless of production levels in other exporting communities. The country's rural sector is far from being competitive and the social conditions of the farmers and their families still need to improve a great deal. For all the historical, economic and social importance coffee carries in Colombia, its fate cannot be left to be determined by temporary and hard-to-predict events affecting other countries. In Colombia, coffee, more than being an agricultural export good, is first and foremost an important component of its social, cultural, institutional and political fabric. It has ultimately served as the basis for the country's democratic stability and national integration (Adjustment Commission for Coffee's Institutionalism, 2002 in Ministerio de Cultura, 2011).

Together, Colombia's coffee stakeholders – from growers to buyers, formal institutions, FNC, co-operatives and farmers themselves – should redouble their efforts to advance and consolidate the country's image as a consistent and reliable producer of coffee of exceptional quality. As this study has shown, private sector companies, using their comparative advantages for innovation, building of human capital and capacity to forge strategic partnerships, can and are already making important contributions to transform coffee production into a more modern, structured, profitable and dignified activity. Nespresso and Nescafé® have shifted production and quality paradigms, encouraged their suppliers to move away from short-term thinking and acting, assisted coffee farmers in differentiating their crops and agricultural practices from those followed in mainstream markets, and paved a quality and value path for coffee producers to deliver taste rather than just delivering beans.

Part of a long process to build trust and loyalty, Nescafé® and Nespresso, coffee producers and the communities they belong to have stopped seeing each other solely as counterparts or disarticulated parts related to each other by mere commercial transactions. With time, they have learned to work as partners. These are powerful reasons to encourage these and like-minded actors to join efforts to build a collective vision and promote sector adaptability, versatility, entrepreneurship and collaboration. Only in this way will coffee production be sustainable in the long run and offer better living conditions to the millions who are part of this social, cultural, economic, political and heritage narrative.

Companies are, first and foremost, interested in being profitable and staying in business over the long term. The challenge and merit resides in redefining the scope of business practices so that these also work in the interest of the society. The result of such shift in corporate models is a socially, environmentally and economically powerful combination. The mix has the potential to build and strengthen communities in a way that concurrently gives private sector firms a competitive edge to remain profitable, minimise deficiencies in the supply chains and continue to thrive over the long term. Under such conditions, development makes good business sense and sustainable business brings about development. The case presented in this book is aimed at illustrating precisely this. Nestlé's engagement in Colombia's coffee sector has responded to market opportunities and sourcing challenges in a manner that acknowledges the brand's interdependent relations with farmers, consumers, shareholders and the community at large. At times, to address the issues that are affecting its supply base, the company has ventured into social, economic, institutional and policy areas.

The absence of and gaps in public policy regarding competitiveness, labour markets, marketing and private sector development urges firms to find ways to overcome structural shortcomings. Many times business has proven to be exemplary, efficient and creative at finding room to manoeuvre and overcome, to the best of their capacity, limits to their growth. However, the very own definition of limits to corporate growth, profitability and commercial success are experiencing an interesting and encouraging transformation and are being increasingly debated. Recently, these issues have gained prominence in the regional development agenda. On 28 October 2013, Nestlé and the Inter-American Development Bank hosted a one-day event for 350 members of the business community, academia, government, social institutions, non-governmental organisations and the society at large to discuss how the private sector can contribute to addressing social challenges and accelerate development in Latin America. Held in Cartagena, Colombia, the fifth Creating Shared Value Forum also touched on how businesses can bring about scalable benefits to its community of interests and help to inform public policies in key productive sectors.

The poverty reduction potential of these public–private synergies is so considerable that Colombia's president, Manuel Santos, urged national and international companies to engrain CSV as part of their long-term growth strategy and actively take part in local development processes (Inter-American Development Bank, 2013a). These types of conversations are key to facilitate the interaction between businesses and the society at large, encourage sustainable private sector investments and reconcile profitability with high development impacts (Inter-American Development Bank, 2013b).

This realisation and acknowledgment is altering the terms and nature of the debate on the links between private sector profitability and their contributions to poverty alleviation and overall societal improvement. International aid agencies and business organisations such as the United Kingdom Department for International Development, the United Nations Commission on the Private Sector and Development, the United Nations Global Compact, the United Nations Environment Programme (UNEP) Finance Initiative, the World Business Council for Sustainable Development and the Inter-American Development Bank (IDB) are already working to harness the best elements of the business world to devise sustainable development solutions (UN Global Compact, 2013; IDS, 2014).

Different sectors of an economic system, a political regime and a society do not have to be at odds with each other. Their progress is certainly not mutually exclusive. Nestlé's activities in Colombia's coffee sector is one such example, as discussed here. Its corporate model, driving operations and management are based on a set of values and purposes that are non-negotiable. It would be disingenuous to believe that shareholders do not expect businesses to remain profitable, returns to be consistent and performance to be competitive.

Yet, it would be equally disingenuous to assume that company owners, also members of society and their communities, are not interested in making that success, growth and competitiveness sustainable and responsible. This is probably a key reason as to why for CSV to be fully engrained in a community, companies endorsing this growth model need to build communication and feedback avenues between the private sector and society. This certainly requires all actors to change their mindsets and the expectations and perceptions they have of each other.

Adopting new concepts, business propositions and private sector engagement strategies require a great deal of thought. Nestlé, like any other company, is confronted with a long list of issues that are yet to be defined and addressed in a rapidly changing world. Terms like 'development' and 'sustainability' are frequently used but are rarely followed by a concise definition. The concept of 'sustainable coffee' remains equally amorphous and elusive. Stakeholders are unlikely to agree on a definition for a term they do not know the exact meaning of. This makes it difficult to set goals, plan activities and estimate for necessary inputs as well as to measure outputs and outcomes to track progress. Social rates of return are yet to be calculated, timeframes to institute meaningful sustainable change are yet to be defined and informative indicators of people's wellbeing are still to be selected.

Firm size thus emerges as an important consideration, especially when defining the scale of responsibility. Ancillary firms, associated distributors, suppliers and stakeholders have somehow come to adopt CSV by proxy. But they too can and should be active and proactive development partners in their communities. Regardless of their size and turnover, they have key roles to play in generating formal employment, reshaping the local labour market and economic landscape and in triggering social development.

After all, small and medium enterprises (SMEs) account for about 90 per cent of businesses and more than 50 per cent of employment worldwide. This, nevertheless, cannot lead to the assumption that SMEs are naturally and intrinsically embedded in their communities. Firm size does not insulate a company from following detrimental practices for their host societies. Yet, the smaller the firm the less public and media scrutiny they face and the fewer resources they have to improve their business processes and practices. It is important these companies are included and encouraged to join the development debate and contribute to the expansion of the traditional 'private sector' definition, which for many scholars and consumers has come to refer solely to large enterprises and multinationals.

The private sector has already taken significant strides in joining communities to create value. This path is expected to continue and widen to include an ever-growing number of sectors, goods, services, raw materials and countries. Certain consumer segments have opened up profitable opportunities by sending market signals that there is an unsatisfied demand for socially and environmentally sustainable products that are also affordable to most and available in mainstream markets. In the past, interest groups and engaged citizens have taken the lead in igniting change. Certified coffee is a powerful example and results have been encouraging. However, and despite its growing importance, it remains a small segment in the global coffee trade.

For sustainability to be successful it should be embedded in every good we consume and every service provided to us. It cannot be a market niche. Society has slowly begun to lead, accompany and help shape this process towards mainstreaming sustainability. Consumers too will have to start taking a long-term perspective with regard to the future availability of agricultural commodities. The most direct way in which they will become engaged is through their consumption patterns. This engagement, however, will have to be more proactive and consistent. Moreover, it will necessarily require a more thorough understanding of the ways governments, local communities and businesses can all work together as development partners, regardless of firm size or commercial activity. In this journey for widespread and affordable sustainability, responsibility, progress and prosperity, society cannot stay last.

Coffee is paving the way to revitalise rural economies, invigorate the agricultural sector and make the supply chain more sustainable. It has become a lead indicator for sustainable crops in terms of social, corporate and civil movements, innovations and practices. For decades, the sector has been engaged in the process of practice fine-tuning and has presented a clear example other commodities can look at as to how this can be achieved. It has also created an extensive organisational, physical, social, cultural, economic and corporate infrastructure supporting coffee growers to become more sustainable and resilient in an ever-changing market. This example is not to be overlooked by other commodity supply chains and industries.

All those stakeholders and shareholders in the coffee sector pushing to mainstream sustainability are reshaping the trade. This is not only enlarging the scope of possibilities farmers have at hand to face an increasingly complex and challenging panorama. Consumers are putting their sustainability needs and wants on the table and making firms respond and anticipate to this. Firms, in turn, are increasingly aware of the linkages grounding them to the societies, economies and ecosystems from where they source their raw materials. Keeping in mind these positive and incremental developments, it is clear that the road towards sustainability and sustainable coffee supply chains is long, demanding and elusive. Sustainability is, after all, not a fixed goal but a continuum of gradually improved relations between people and its surroundings.

Ultimately, what the agricultural sector is in urgent need of are farming practices that lead to greater productivity and increased quality and partners that can link them to domestic and global markets that pay for higher quality raw materials. Private standards and certification schemes have done a great deal of good in this regard. Yet, they are far from being panaceas to all systemic challenges and sudden contingencies. They cannot offer exhaustive solutions to sustainability and production challenges at the macro level, halt environmental degradation in coffee producing regions all together and alleviate social tensions at the community or family level. What they can do, and have done, is open up new opportunities for farmers. Bacon et al. (2008) applaud these achievements. They also suggest it is due time to think critically about the ability these sustainability-oriented standards have to deliver on their goals. How else will these initiatives realise the promised potential to contribute to making farmers' livelihoods more secure?

Supporters and critics alike should reasonably determine what standards could realistically achieve. Perceived as sustainability successes, standards, verifications and certifications point to a far more important task. That of establishing roles and cooperation modalities for public, private, social, academic and sectoral actors to forge strong partnerships and provide the sort of effective and long-term support hard-pressed coffee farmers require to overcome their current precarious socio-economic conditions (Baker, 2013; Bacon et al., 2008). After all, umbrella-wide guidelines alone can only begin to accommodate and reflect the extremely broad range of local realities shaping coffee production. These are powerful reasons explaining why coffee is seen as the fire test for sustainability enthusiasts, globalisation defenders and corporate leaders alike.

Despite their many hurdles and struggles, for many farmers in Colombia, and other producing countries, coffee will continue to be their source of livelihood, the engine behind economic development in their communities, the entry point for technological upgrading, and a force for social dynamism and environmental conservation. As long as coffee remains the drink of choice for millions of consumers around the world, coffee can successfully become one of the world's most powerful tools for social change.

Postscript

In the last days of October 2014, as the manuscript for this book was being finalised for publication, Colombia's coffee sector became involved in quite a stir. The agitation was not directly related to Nestlé and its sustainability-oriented schemes but had a lot to do with the institutional framework that regulates coffee production in Colombia. The culprit was a government-commissioned study tasked to make contributions to the sustainable and competitive development of Colombian coffee in the medium and long term.

The study, involved actors, its findings and elicited reactions constitute a heterogeneous mixture of opinions and interests. The whole episode ultimately reflects the existence of numerous factors and forces in the sector and the opposing directions in which each one of them is pulling. The sector is complex and so will be the process of making it more competitive. This is after all, the shared C in Colombia and coffee discussed in an earlier chapter: complexity. Judging by the immediate response to the study findings, the report seemed to have pleased few.

Through the National Council for Economic and Social Policy (CONPES), President Juan Manuel Santos established in 2013 a Mission Study for the Competitiveness of Coffee in Colombia. Led by renowned economist Dr Juan José Echavarría, the Mission was given the task to conduct a thorough and comprehensive analysis of the coffee sector in Colombia and the world. With the findings, the team had to design a set of public policies and strategies to address structural problems. Particular attention was to be paid to issues related to production, marketing, innovation, value-added, risk management, social aspects such as employment and income, as well as to the country's coffee institutional framework.

A great deal of controversy emerged from the premature and unauthorised release of the draft report. To a certain extent, findings and reactions to them were not large surprises. Productivity was singled out as the most pressing challenge to tackle. This was a predictable finding. For more than 20 years, productivity levels have remained unchanged. Yields per hectare in 2014 stood at a low 16 bags per hectare, which meant that even if prices were and remain high, coffee production is still not a profitable activity.

Amongst other things, the document proposed a series of reforms to the FNC, starting by the separation of commercial and management issues. This can barely constitute an unexpected recommendation. The response could have barely surprised anyone. The FNC was quick to express its disagreement and dismiss the findings. Authors were fiercely accused of wanting to dismantle the institutional architecture that offers public services such as the purchase guarantee, agricultural and technical support, and promotes the differentiated quality and value of Colombian coffee.

The FNC has a robust history in the country that is more than 90 years strong. It is normal the organisation wishes to protect and preserve the institutional architecture and governance structure it has built over those decades. But coffee growers need more than history and past achievements to overcome systemic challenges in coffee production. Since current and past measures have not proven successful in making coffee production sustainable and profitable, maybe it is time to make more profound reforms to the institutional framework regulating the coffee trade in Colombia. This could have only appeared as a radical measure to an institution that has looked too much and too often at the past.

This book repeatedly stated how the coffee world has changed. But domestic institutions have not been forthcoming of this new scenario nor have they adapted to it. Failing to adjust to new market conditions after the end of the quota system, Colombia gradually lost ground in the global market for Arabicas. This trend was only halted in 2012. However, the number and severity of challenges farmers have to face remain as before. Opportunities have also been missed. For example, the Mission regrets Colombia's absence in the growing market for Robusta. Ten years ago, Robusta beans constituted around one-third of world demand. This has now increased to around 40 per cent. In addition, the study stressed the very recent penetration of the country's beans in niche market for speciality coffees despite tight controls on the quality of coffee available for export.

Taking a close look at Echavarría's Mission findings, there are areas of convergence with some of the arguments presented in this book. Subsidies, policies and institutional reforms are only some of them. We briefly mentioned how subsidies have prevailed over the delivery of public services. Handouts, as opposed to the extended provision of public goods and infrastructure, constitute one of the country's main strategies to promote growth and development in agriculture and more notably in coffee. Moreover, previous sections expressed a fair degree of reservation regarding the overlapping roles the Federation plays as a public and private entity.

The FNC acts as a private exporter, public regulator and trade union. It also assumes public functions, drafting policy as well as managing coffee contributions, public funds and the National Coffee Fund. As an exporter, and unlike its competitors, the FNC incurs no fiscal burdens nor faces the same risks. Financial deficits are closed with public resources. Interests were bound to conflict.

This book has made reference to the national policy for rural development as an invaluable opportunity to promote endogenous growth in the countryside. One way it could do this is by taking a more comprehensive approach to growth. Development in rural and urban areas is not an independent process. Since they feed into each other, policies will meet limited success if linkages connecting them are overlooked. Even within the rural sector, there are numerous productive linkages that should also be explored.

For instance, agriculture is the foundation of rural economies and communities. Its importance is unquestionable but does not rule out the opportunity to promote other productive activities. The same holds true for coffee production. In many regions, coffee is at the heart of local agricultural economies and communities. For decades, this centrality in and to rural and agricultural life in Colombia has separated coffee development and social policy from public agricultural policies and rural development schemes. Strategies and action plans have been left for the coffee growers' union to design.

Present realities, however, suggest the government should take the lead in formulating policies in the coffee sector, making it part of the overall framework to generate growth in agriculture, the rural sector and in the country. After all the State is responsible, and accountable, for delivering public services, goods and infrastructure, alleviating poverty and formulating public policies. The Mission Study for the Competitiveness of Coffee in Colombia argues in favour of analogous measures and some more.

Productive development policies and social policies have been fused. In reality, these are policy strands with very distinct objectives. The Mission unequivocally advocates for a series of reforms to reorganise the sector. It argues institutional changes could open up opportunities for the private sector to take part in coffee production and commercialisation on the same grounds as the FNC. They would also dissipate conflicts of interest in the multiple roles assumed by the FNC as sector regulator, policy implementer, producers' representative and exporter. Lastly, institutional reform would also separate the public and private management of resources pooled in the National Coffee Fund.

The draft report calls for the responsibility to design coffee policy to be given to the government. In that way, and similar to other private actor or producers' unions, the FNC would advocate for the needs of the group it represents before the relevant state bodies. The government would also have to be in charge for formulating and executing social policies directed at coffee growers. As a result, the reformed FNC would continue to exist as a union and player in the coffee market, albeit from a position where it would face more competition. Coffee contributions would be used to fund research & development activities and technical assistance.

Deregulating coffee exports is another of the reforms that has been deemed as indispensable. The authors sustain that the freer the trade and the less the government intervenes, the more efficient the production of coffee will be. Currently, the Federation simultaneously regulates and competes with other private exporters to take Colombian coffee abroad. In fact, the FNC has traditionally kept a tight control on the quality of Colombian coffee that makes it to the global market. The best beans are exported. The world thus enjoys the best of Colombia's coffee crop. Lower grades are kept for domestic consumption.

The Mission urges this situation to change. Quality standards should not be used as export barriers. Coffee farmers failing to produce *excelso* and *supremo* beans should be allowed the opportunity to place their crops in international markets. If demand exists for these coffee grades, there is no reason to exclude them from or prevent their entry into global supply chains.

More nuanced recommendations have to do with the power the FNC has to set domestic coffee prices. According to both the Federation and the popular movement Dignidad Cafera, this is no insignificant safety mechanism for small farmers. Without price support, many coffee producers would be at the mercy of the interests of large corporations. In 2014, some 800,000 individual purchases below 25 kg were made. This is used as strong evidence that small producers are the real beneficiaries of this price control mechanism. Growers evidently favour the existence and continuity of the purchase guarantee. It allows them to secure a minimum price, a fixed buyer and cash payments for their crops that are closer to international prices.

However, there are powerful reasons to question the effectiveness and unintended uses of support measures. The report's main author, Juan José Echavarría, argued that not enough information about production costs was available when the subsidy scheme to protect coffee earnings (*Protección del Ingreso Cafetero*, PIC) was calculated. This would explain why 60 per cent of the allocated resources went to 10 per cent of all producers. The PIC has not inevitably brought about benefits, in the expected magnitudes, to the intended groups. The Mission Study suggested something similar has happened to support mechanisms like purchase guarantee. There is not strong enough evidence to conclude that the purchase guarantee is effectively transferring a greater price share to farmers. Compared to other coffee producing countries that do not have purchase guarantees, Colombia does not fare better. Finding ways it does motivated the commissioning of this study on the competitiveness of coffee in Colombia.

Despite and even more so because of all the controversy, the aim of the Mission Study is to find solutions to boost competitiveness. The FNC reaction has been strong. So has been the response by other groups. When scrutinised, it is normal for individuals and institutions to succumb to the temptation of dismissing feedback. Nevertheless, if history is to give any cue about the future, market conditions cannot be expected to remain forever favourable to Colombian farmers.

Many times in the past have cyclical forces shaken the sector and exposed the structural fissures that hamper growth. In fact, commodity prices have been more volatile in the last ten years than in the previous past. It is thus too soon for any stakeholder, public or private, to dismiss the study and its findings. Some of the proposed measures are contentious. After all, they are calling for major adjustments to be made in policies and institutions that may meet resistance. Opinions are naturally going to differ.

Difference may be just what Colombia needs. Different points of view, different voices, different options and different policies. A marked shift from how the sector has worked in the past. The authors wholeheartedly hope that the Mission's findings are painstakingly analysed. That may just open up a nationwide forum to conduct a frank and inclusive dialogue to formulate the sort of public policies the coffee sector in Colombia veritably needs.

Epilogue

Peter Brabeck-Letmathe, Chairman, and
Paul Bulcke, CEO, Nestlé SA

The true test of a business is whether it creates not only shareholder value but also value for society over the long term. As a company, Nestlé is all about quality of life and nutrition. It delivers products and services that offer people ways towards improved nutrition, health and wellness. Its objective is also to be the industry reference for financial performance, trusted by all stakeholders. Its ambition is its motto: 'Good Food, Good Life'.

Consistently delivering on our promises and building trust is the only way we can achieve this objective. We realise business can prosper over the long term if, and insofar, as the communities we serve also prosper. We call this approach Creating Shared Value. This premise is at the heart of the strategy we have followed to build our brand value, position ourselves ahead of the curve and drive competitive advantage through innovation and research & development (R&D). Our coffee products, many of which are discussed in this book, perfectly embody our innovative spirit and our commitment to build and reinforce in the long-term interest of the society in our business case.

Our first ever product was the result of innovation and the drive our founder, Henri Nestlé, had to address a latent social need: decreasing malnutrition and develop a life-saving infant cereal. Today, this commitment to offer consumers healthier natural food and beverages at all stages of life is stronger than ever and is extended to science based nutrition and dermatology health solutions. Our underlying CSV approach has been initially concentrated on three areas: Nutrition, Water and Rural Development.

Good nutrition will continue to play an ever more important role in the health and wellness of individuals and society. In 2013 alone, we renovated 7,780 products because of nutrition or health considerations and increased the nutritious ingredients and essential nutrients in 4,778 products. Our focus on water comes from our understanding that water is quite simply the linchpin of food security. As water scarcity becomes an even more serious and widespread issue in many countries, it may compromise food security.

Our focus on rural development is because suppliers, farmers, rural communities and small entrepreneurs are intrinsic to the long-term success of our business. Our long-term ability to source the right quality and quantity of our key inputs depends on the sustainability of farming all around the world.

We continue to further our Farmer Connect programme to buy directly from producers, co-operatives and selected traders who apply Nescafé®'s Better Farming Practices and receive free technical assistance and training. This sourcing concept brings us the advantage of a shorter supply chain, lower transaction costs, a direct influence on quality and a deeper understanding of the challenges and opportunities our farmers face. As part of our rural development and responsible sourcing activities, we have trained 300,000 farmers all around the world.

While concrete action plans were executed in the three areas, Nestlé has sought to distinguish itself from the broader debate on corporate social responsibility and stimulate further discussion on Creating Shared Value. Seven years after we defined the three areas of focus for our shared value creation strategy, we realise we have come a long way but still have a long way to go.

In 2012, we shared with the public a series of robust commitments to support our long-term goal of Creating Shared Value. They cover nutrition, health and wellness, rural development and responsible sourcing, water, environmental sustainability, our people, human rights and compliance. One year later, we added ten new commitments and adopted short-term objectives for each one. By sharing our goals externally and publicly, we are making it possible for stakeholders to hold us accountable for our achievements and challenges.

Our past efforts to Creating Shared Value have taken us very far but we are aware that we need to become more systematic and structured in the way we proceed to implement this business proposition. We still have to fully implement the framework that guides our efforts in maximising the shared value we create for our business and society. We also know we must take a long-term view for a company like ours to prosper. For example, the corporate business principles and management and leadership principles that guide our actions have been developed over close to 150 years.

The opportunities and challenges ahead are many, especially as we work to create shared value at every step of our value chain for each one of the more than 10,000 trusted products we manufacture, and in each of the 196 countries where we have sales operations. For this to be possible, our company will need to maintain itself at the forefront of technological, technical and partnership innovations.

Over the decades, we have built a series of elements that contribute to the company's competitive advantage. Our product and brand portfolio is unmatched thanks to our R&D capability, wide geographic presence and the entrepreneurial and innovation-driven spirits of the people who make Nestlé. We have 5,000 people working in 34 R&D facilities around the world, which makes us the largest R&D global network in the food and beverage industry. One of the operational pillars that have positioned Nestlé at the forefront of product development, renewal and consistent quality is precisely innovation.

As noted earlier, Nestlé's history stretches back nearly 150 years and has been characterised by a forward thinking, pioneering approach. We ventured out of our home country, Switzerland, and established factories in places where we have continued to invest on the foundations for long-term growth: people's skills and capabilities and a network of local partners. Over the course of many decades, and in all the countries where we operate, we have worked to create a more conducive business environment, improve the capabilities of farmers to be more productive and deliver higher quality inputs, improve the skills of our workforce and develop and implement standards. This has substantially increased our understanding of how we can contribute to society in a way that reinforces our business model.

At each step of the value chain, our actions can potentially produce social and environmental benefits or result in negative impacts. This is particularly true for milk, cocoa and coffee, for which we are the world's largest purveyors. For instance, Nestlé buys 10 per cent of the world's production of coffee.

This means we are in a privileged position to advance our belief that in order to prosper we need the communities we serve and in which we operate to prosper as well. This is why we need to work relentlessly to fulfil our corporate responsibilities and tie local needs and opportunities with our business objectives. Working with and for society has ultimately built the basis for our sustained growth in local markets and also worldwide.

Coffee, the issue discussed in this book, provides an excellent example of how consumers, society at large, development agencies, NGOs, public institutions and Nestlé have contributed to improving the lives of thousands of coffee growers in Colombia, probably the most recognised coffee producing country in the world. Even when we are only one of the many players shaping the country's coffee sector, we are committed to bringing prosperity to the farmers who supply us through improvements in our value chain. The popularity of our coffee brands also attest to the many innovations the company has brought to the industry.

Nescafé® is the world's favourite and largest coffee brand. It is both a global icon and a local favourite sold in almost every country in the world. Some 5,500 cups of Nescafé® are drunk every second of every day. Launched in Switzerland in 1938, the brand came as a solution to an oversupply of green coffee in Brazil and the risk of having to throw those extra green beans to waste. At that time, Nestlé saved numerous farmers from bankruptcy, and now the brand is one of the most successful in our company. Forty-five years later, the worldwide coffee crisis urged our specialty-coffee brand, Nespresso, to launch a programme to assist farmers to raise productivity and quality, the main determinants of the incomes they derive from coffee production.

Similarly, Nespresso is in itself an innovative breakthrough. It created a new category of coffee altogether, the capsule coffee market, and its sourcing strategy has focused on the most important factors determining farmers' income: quality, productivity and sustainability. The Nespresso AAA Sustainable Quality™ Program, discussed at length in the book, was launched in November 2003 at a time when farmers were facing considerable challenges to continue production. Already at the time, specialisation was seen as an example of the alternatives available to farmers to overcome the crisis in the sector. The technical assistance and premiums the AAA Program began offering farmers have since then had positive impacts on their livelihoods. Not only are they producing highest quality coffee and increasing yields, but also independent studies show that the coffee growers selling their beans to Nespresso live in healthier environments, use natural resources more sustainably and are able to lead more prosperous lives.

To continuously develop new coffee products and renew popular brands, we maintain our commitment to source green beans directly from the farmers, assist growers in raising productivity, build and invest in processing factories in coffee producing countries themselves, and integrate our activities with the local economies and communities. This requires innovation at the corporate, strategic, partnership, policy and implementation levels. Nestlé is investing CHF 500 million in coffee projects over the decade 2010–2020. This includes CHF 350 million for the Nescafé Plan and CHF 150 million for Nespresso.

The success of both Nescafé® and Nespresso depends on two fundamental elements: our suppliers and the quality of the crops they produce. We know crop quality may vary from batch to batch. Yet, our products have to be consistent and deliver to our consumers the same taste and experience they are expecting every time they open a jar or sachet of Nescafé® or introduce a capsule of Nespresso into their machines. For product consistency and brand growth, we need to improve the quantity and quality of green coffee we source from around the world. We also need to make sure that supplying farmers are producing the crop sustainably and without compromising the natural resources upon which their present and future livelihoods depend.

The Nescafé Plan and AAA Nespresso Plan have distinctive features, standards and requirements, reflecting their different business models and sourcing operations. Yet, they are two important initiatives by which we are seeking to increase the quality of the beans we purchase and pass on these quality gains to our consumers. In the process, we are assisting farmers almost in the entire coffee growing belt and particularly in Colombia, Costa Rica, Sudan, Mexico, Kenya and other important coffee growing countries to adopt better farming techniques. This support helps them to increase yields, productivity and output of higher quality beans. Better quality raw materials ensure higher prices and thus higher incomes for farmers. In the future, we expect our farmers to continue to acquire more and better skills, adopt more sustainable agricultural practices and become increasingly more prosperous.

The quality and taste of every cup of Nescafé® is the result of continuous technical innovations. Coffee and cocoa are important crops for Nestlé. They are the source of livelihoods for hundreds of millions of farmers around the world. Yet, they count among the forgotten species of the seed industry. This is why Nestlé invested significantly in coffee seed research at the Nestlé R&D centre in Tours. Its purpose is to improve cocoa and coffee's agronomic traits, ensure better field yields and disease resistance, improve processing characteristics, and enhance the taste of products. Together with good farming practices, such progress can help farmers improve production and increase their income.

Nestlé's beverage R&D capabilities cover all aspects from farm to cup, including raw materials, flavour extraction, systems, packaging and retail channels. Innovation has been at the heart of what is now a heritage and tradition. A part of it is visible to our consumers, the other one not so much. For example, the development of the freeze-drying technique led to the introduction of Taster's Choice instant coffee in 1966. Later, the brand was the first one to launch a soluble coffee-based cappuccino and latte macchiato, the breakthrough system Nescafé Dolce Gusto, and a series of 3 in 1 varieties to take the brand to countries with no coffee drinking tradition. Lately, we introduced Nescafé Smoovlatté in China, opening up the ready-to-drink segment in the country.

After 75 years, the brand has reinvented itself to offer the same sense of adventure it did when it was first launched. Nescafé® accompanied the first successful expedition to climb Mount Everest and astronauts to the moon. Now, to maintain the brand freshness and reach out to new consumers, with different lifestyles and tastes, Nestlé has opened a new R&D centre in Singapore. It will take the global lead for the company's innovations in Nescafé® coffee mixes and other leading powdered beverages like Milo. Nutrition and health are integral in new product development and so the resulting developments will be reflected in products that meet local tastes with less sugar, salt and fat. In this way, Nestlé's global brands become local products in each of the 150 countries where our products are available.

We are looking to promote responsible farming, production and consumption to satisfy a growing demand for coffee. This is why responsible sourcing is key. Smallholders, who produce most of the coffee in the world, farm in plots no bigger than 2 ha. They also face volatile prices, declining yields from ageing trees and/or plant diseases, climate change, poor infrastructure, crop substitution, rural outmigration, among others. If they do not produce enough good quality coffee, the company cannot secure the inputs it needs for Nescafé®, Nespresso and Dolce Gusto® to maintain their position as the world's leading coffee brands. If our farmers prosper, so will Nestlé.

Coffee production in Colombia, a key coffee procurement market, faces similar challenges. This book analyses the problems the farmers are facing in depth. But we are making progress in addressing some of the structural factors that limit productivity and quality and hamper the improvement of farmers' lives through the AAA Program and Nescafé Plan. Our goals are ambitious.

Through the Nescafé Plan, we are investing CHF 500 million in coffee projects over a period of ten years to breed and distribute 220 million high yield, disease resistant coffee plantlets by 2020, train farmers according to the Nescafé Better Farming Practices, as well as apply the fundamental sustainability criteria. The AAA Sustainable Quality™ Program launched in August 2014 its sustainability strategy for 2020: The Positive Cup. Among other objectives, the premium brand is committed to sourcing 100 per cent of its coffee sustainably and to assist farmers to achieve high certification standards in partnership with the Rainforest Alliance and Fairtrade. In Colombia we have partnered up with key local institutions, such as the Coffee Growers Federation in Colombia, to assist farmers in building a robust business.

Innovations also extend to the activities the brand implements to increase farmer welfare and attract young farmers to remain in coffee production. Nespresso will expand the AAA Farmer Future Program that first started an initiative to create a retirement fund for farmers in Colombia.

We want to improve the overall wellbeing of working and future coffee growers, rural communities, small entrepreneurs and suppliers. However, the economic, social and environmental challenges they are confronted with cannot be tackled by one organisation alone. This is why we believe in establishing long-term, mutually beneficial partnerships where each stakeholder can contribute to achieving shared goals by offering its expertise. The production of our coffees offers excellent examples of the collaborative initiatives that have been established to connect our business to local and global partners.

Cross-sector partnerships, in which we marry our technical expertise and research in crop optimisation with our partners' local community knowledge, are helping us to improve our strategies. This is why we have established partnerships with actors who share our commitments to sustainability and make our goals credible and deliverable. We believe in working collectively at the global and local levels through partnerships and multi-stakeholder platforms, including UN agencies, other international organisations, governments, academia and NGOs. This allows us to listen to and learn from different opinions, share experiences and further our implementation of good practices.

The list of organisations we work with is extensive. Coffee is a key input for us and so we have established partnerships in all our supplying countries to find sustainable suppliers by creating long-term, viable supply chains and enabling producers to get a better price for their products. We also engage with global, regional and local platforms to help address common industry challenges. This allows us to seek opportunities for non-competitive solutions that will achieve positive change.

Through our Nescafé Plan we continue to drive adoption of the Common Code for the Coffee Community (4C Association) by farmers supplying coffee to us. Adoption of this verification system contributes to the improvement of economic, social and environmental conditions of those who make their living from coffee. We are also part of the Sustainable Coffee Programme, an IDH-facilitated pre-competitive initiative in which public and private actors have come together to expand sustainable coffee production into mainstream markets. This platform aims to help millions of coffee farmers become more resilient and to implement social and environmentally friendly coffee-farming techniques to increase yield and quality.

Through effective alliances, we offer a more efficient route to market, local training, plant propagation and distribution, and technical assistance from our team of 220 agronomists. In 2013, this team of sourcing staff visited more than 30,000 coffee farmers in 14 countries to provide expertise and help to implement good practices in the field.

In Nestlé we are aware that innovation and sustainability also come from how we manufacture our products. In the coffee processing plant of Bugalagrande in Colombia, 90.2 per cent of the generated solid wastes, some 16,000 tonnes, come from cisco, the husk left from the coffee production process. In 2009, we invested in the construction of a biomass boiler to convert coffee husk into energy, the first plant of its kind on the continent. Coffee husks are now meeting one-third of the plant's energy requirements. The cisco plant has contributed to energy savings for the factory, ensured cleaner production processes and reduced the amount of solid waste going to the local landfills.

Beyond making the most innovative products in the most efficient way, we also need to ensure that our products are available sustainably whenever, wherever and however consumers want them – and all along the consumer spectrum. In Central and West Africa we launched the MYOWBU (My Own Business) initiative. We trained entrepreneurs to treat street selling as their own business and they made Nescafe® available within arm's reach to consumers out of home – and in the streets.

This programme helps previously unemployed young people create their own business selling Nescafé® coffee. MYOWBU quickly expanded from the original countries of Burkina Faso, Cameroon, Côte d'Ivoire, Ghana, Nigeria and Senegal to South Africa, Angola, Mozambique, Kenya, Zambia, Democratic Republic of Congo, as well as in the Maghreb region. It has created close to 3,800 jobs and has sold 70 million cups of Nescafé®. We are concerned about the high youth and women unemployment in many countries, not only in emerging markets. Through the MYOWBU and the Nescafé Street Barista campaign in Thailand, we have reached more than 600,000 women.

In Europe, youth unemployment has become a major issue with one in four young Europeans – about 5.6 million people – affected. It is also a growing market for our company and we need to prepare the next generation of professionals. To help to address the unemployment crisis and attract and retain some of the best talent, Nestlé has offered thousands of jobs, traineeships and apprenticeships to young people across Europe. In Girona, Spain, one of the worst affected countries, we have expanded our Nescafé Dolce Gusto plant, generating 150 new jobs since it was opened in 2011.

We have also developed unique retail concepts. We have taken Nespresso to new premium locations, for instance airports. At the Barcelona Airport, travellers can now purchase Nespresso's Grands Crus or Limited Edition from an automated Cube. This is the latest addition to the already unique retail profile offered by Nespresso: 320 retail boutiques worldwide, consumer relationship centres, the online boutiques at www.nespresso.com and now also by the Nespresso Cube.

As a company, we are still working on setting commitments, targets and roadmaps to lay out a clearer strategy of where we are heading to and the standards we uphold and to which we hold ourselves accountable. Nestlé has developed performance indicators to provide a focus for measuring and reporting Creating Shared Value, sustainability and compliance. This performance summary forms part of our communication on progress regarding the United Nations Global Compact Principles, a global strategic policy initiative for businesses committed to aligning their operations with ten universally accepted principles in the areas of human rights, labour, environment and governance.

Our partners and external organisations also help us measure our performance, give us feedback and contribute to the improvement of our strategies. Our consumers also have a key role to play in this. Many brands are market leaders and have built a powerful consumer loyalty. We must do a better job of informing consumers and society at large that they can be confident of the social benefits they are helping create by drinking a cup of coffee from one of our leading brands.

Effective dialogue with our stakeholders is central to Creating Shared Value. At the global level, we do this through the Creating Shared Value Forum series. These dialogues allow us to engage with experts from government, academia, civil society and business to discuss how to accelerate sustainable development. On 28 October 2013 we co-hosted, in collaboration with the Inter-American Development Bank, the fifth Creating Shared Value Forum in Cartagena, Colombia. On that occasion we held discussion on 'Creating Shared Value: The changing role of business in development'. The Colombian President, Juan Manuel Santos, opened the event and outlined the role of the private sector in the economic and social development of Colombia.

To us, this was an important sign that governments and private sector companies are beginning to work closer to each other to bring benefits to society. We recognise that our position in society brings both opportunities and responsibilities. But Nestlé is, after all, only one of the many companies in the food and beverages sector and in the world looking to create and implement innovative solutions to the world's most pressing challenges. Only if we work with key external and internal stakeholder groups will we be better prepared to face the many challenges ahead and leverage the enormous scale and reach of the private sector to help solve social problems through our core business.

We are committed to Creating Shared Value. This is the way we do business and the way we connect with society at large.

Bibliography

All websites were available at the time of writing.

4C Association (2010). *FNC: Portrait of a Coffee Federation*. Retrieved from: www.4c-coffeeassociation.org/uploads/media/4C_Portrait_FNCJan_2010_EN.pdf?PHPSESSID=hu8mlb9ljm0qerfk1paelmc537

4C Association (2012). *Essentials*. Retrieved from: www.4c-coffeeassociation.org/uploads/media/4CDoc_001_4C_Essentials_v2.2_en.pdf

Aeropuerto del Café (2014). *Reseña Histórica Del Aeropuerto Del Café*. Retrieved from: www.aeropuertodelcafe.com.co

Agricultural Cooperative Development International (ACDI) and Volunteers in Overseas Cooperative Assistance (VOCA) (2014). *Colombia-USAID Specialty Coffee Program*. Retrieved from: www.acdivoca.org/site/ID/colombiaCAFES/

Aguadas in Caldas (2014). *Hoy en Aguadas, Caldas MintrabajoCol y Nestle-Nespresso firman Alianza Público Privada que Permitirá a Caficultores Ahorrar para su Vejez*. Retrieved from: www.aguadas-caldas.gov.co/index.shtml#6

Akiyama, T. (2001). Coffee market liberalisation since 1990. In T. Akiyma, J. Baffes, D. Larson and B. Varangis, *Commodity Markets Reforms: Lessons of Two Decades*. Washington D.C.: World Bank Publications.

Almacafé (2010). *Quienes Somos*. Retrieved from: www.almacafe.com.co/es/quienes_somos/almacafe_sa/

Aluminium Stewardship Initiative (ASI) (2014). *ASI Standard Draft0: Synopsis of the Comments Received during the First Public Consultation (1 February to 29 March 2014)*. Retrieved from: http://aluminium-stewardship.org/wp-content/uploads/2014/04/ASI_Synopsis_1st-Public-Consultation.pdf

Alvarado-Alvarado, G., Posada-Suárez, H. E. and Cortina-Guerrero, H. A. (2005). *CASTILLO: Nueva Variedad de Café con Resistencia a la Roya*. Avances Técnicos. Manizales: Cenicafé.

Alvarez, G. and Wilding, R. (2008). *Governance Dynamics in a Multi-stakeholder Network: The Case of Nespresso AAA Sustainable Quality Program*. Conference Paper, British Academy of Management.

Ananthanarayanan, R. (2014). *Coffee Prices Soar on Dry Weather in Brazil*. 1 March. Retrieved from: www.livemint.com/Money/NtQunNI7PzMrnsEI9jy5CI/Coffee-prices-soar-on-dry-weather-in-Brazil.html

Anderson, K. and Valdés A. (eds) (2008). *Distortions to Agricultural Incentives in Latin America*. Washington DC: World Bank.

Anderson, W. and Vandervoort, C. (1982). *Rural Roads Evaluation Summary Report*. Washington, DC: United States Agency for International Development.

Andrade, R., Overby, D., Rice, J. and Weisz, S. (2013). *Coffee in Colombia: Waking Up to an Opportunity*. 2 January. Retrieved from: http://knowledge.wharton.upenn.edu/article/coffee-in-colombia-waking-up-to-an-opportunity/

Ángel, F. (2013) *Is Colombia Suffering the Dutch Disease?* Retrieved from: www.huffingtonpost.com/felipe-angel/is-colombia-suffering-the_b_3247265.html

Areas for Municipal Level Alternative Development Program (n.d.). *Colombia: Areas for Municipal Level Alternative Development Program (ADAM)*. Retrieved from: www.tetratechintdev.com/index.php?option=com_k2&view=item...pal-level-alternative-development-program-adam&Itemid=60&lang=us

Aristizábal, C. and Duque, H. (2008a). Identificación de los patrones de consumo en fincas de economía campesina de la Zona Central Cafetera de Colombia. *Cenicafé*, 59, 321–342.

Aristizábal, C. and Duque, H. (2008b). Identificación de los patrones de ingreso en fincas de economía campesina de la Zona Central Cafetera de Colombia. *Cenicafé*, 59, 358–375.

Arpal (n.d.). *CELAA, Una Iniciativa Francesa Para Recuperar Envases Metálicos Ligeros*. Retrieved from: http://aluminio.org/?p=1283

Asociación de Banco de Alimentos de Colombia (ABACO) (2013). *Bancos de Alimentos de Bogotá*. Retrieved from: www.abaco.org.co/index.php?option=com_content&view=article&id=146&Itemid=63

Asociación Nacional de Anunciantes (2013). Manuel Andrés ¿Por Qué Celebra Nestlé? *Revista Anda*, No. 50.

Asociación Nacional de Anunciantes de Colombia/National Association of Colombian Advertisers (Andacol) (2013). ¿Por qué Celebra Nestlé? *Revista ANDA*, No. 50, 16 April.

Asociación Nacional de Café (de Guatemala) (Anacafé) (2014). *Precios del Café*. 28 February. Retrieved from: www.anacafe.org/glifos/index.php?title=Especial:GraficaDePreciosDelCafe

Asociación Nacional de Empresarios de Colombia/National Business Association of Colombia (ANDI) (2010). *ANDI Premiará la Innovación de los Proveedores*. Retrieved from: www.andi.com.co/pages/noticias/noticia_detalle.aspx?IdNews=107

Asociación Nacional de Exportadores de Café de Colombia (Asoexport) (2014). *Colombia enfrenta escases de recolectores de Café*. 13 September. Retrieved from: www.asoexport.org/Colombia-enfrenta-escases-de-recolectores-de-cafe

Auld, G. (2010). Assessing certification as governance: effects and broader consequences for coffee. *The Journal of Environment & Development*, 19(2), 215–241.

Azahar Coffee (2013). *Coffee Varietals*. Retrieved from: https://azaharcoffee.com/varietals/variedad-colombia-castillo

Bacon, C. (2005). Confronting the coffee crisis: can fair trade, organic, and specialty coffees reduce small-scale farmer vulnerability in northern Nicaragua? *World Development*, 33(3), 497–511.

Bacon, C. M., Ernesto Mendez, V., Gomez, M. E. F., Stuart, D. and Flores, S. R. D. (2008). Are sustainable coffee certifications enough to secure farmer livelihoods? The millennium development goals and Nicaragua's Fair Trade cooperatives. *Globalizations*, 5(2), 259–274.

Baker, P. S. (2013). The changing climate for sustainable coffee. In Proceedings of the 24th International Conference on Coffee Science (pp. 603–610).

Balch, O. (2011). "Sustainable Agriculture Briefing Part 2: Corporate practices - Fixing the first link in the chain" Retrieved from: www.ethicalcorp.com/communications-reporting/sustainable-agriculture-briefing-part-2-corporate-practices-fixing-first-li

BareFoot Foundation (2008). *A Long Term Treaty. Nestlé and the Pies Descalzos Foundation.* Retrieved from: www.fundacionpiesdescalzos.com/barefoot-fundation/pages/news/news_main_en.php

Bejarano, J. A. (1987). *El Despegue Cafetero (1900–1928).* In J. A. Ocampo (ed.) *Historia Económica de Colombia* (pp. 195–227). Bogotá: Siglo XXI, Editores de Colombia.

Benavides, J. (2003). *Infraestructura y Pobreza Rural: Coordinación de Políticas e Intervenciones en Países de América Latina y el Caribe.* Working Paper. Washington, DC: Inter-American Development Bank.

Bennett, J. (2002). Multinational corporations, social responsibility and conflict. *Journal of International Affairs-Columbia University, 55*(2), 393–414.

Bergquist, C. (1986). *Coffee and Conflict in Colombia, 1886–1910.* Durham, NC: Duke University Press.

Bertrand, B., Boulanger, R., Dussert, S., Ribeyre, F., Berthiot, L., Descroix, F. and Joët, T. (2012). Climatic factors directly impact the volatile organic compound fingerprint in green Arabica coffee bean as well as coffee beverage quality. *Food Chemistry, 135*(4), 2575–2583.

Biswas, A. K., Tortajada, C., Biswas-Tortajada, A., Joshi, Y. K. and Gupta, A. (2014). *Creating Shared Value: Impacts of Nestlé in Moga, India.* Berlin: Springer International Publishing.

Bitzer, V., Francken, M. and Glasbergen, P. (2008). Intersectoral partnerships for a sustainable coffee chain: really addressing sustainability or just picking (coffee) cherries? *Global Environmental Change, 18*(2), 271–284.

Bloomberg (2013). *Markets, Company Quotes: NESN: VX.* Retrieved from: www.bloomberg.com/quote/NESN:VX

Bloomberg (2014). *Foreign Exchange Cross Rates.* Retrieved from: www.bloomberg.com

Botello, S. (2010). Jornales Cafeteros e Integración del Mercado Laboral Cafetero: 1940–2005. *Ensayos sobre Economía Cafetera,* No. 26.

Bragg, E., Krogseng, K. and Schwaller, C. (2013). *Leveraging a more sustainable global agricultural system: improving multinational organizations' capacities to procure.* (Master's thesis). Retrieved from: www.bth.se/fou/cuppsats.nsf/all/67b0f12526338becc1257b8d00591e74/$file/BTH 2013 Bragg.pdf

Brando, C. (2012). *Brasil, País del Café: Oportunidades y Amenzas.* Espírito Santo do Pinhal, Brasil: P&A Marketing International.

Bright, B. (2008). How more companies are embracing socially responsibility as good business. *The Wall Street Journal,* 10 March. Retrieved from: http://online.wsj.com/news/articles/SB120491426245620011?mg=reno64-wsj&url=http per cent3A per cent2F per cent2Fonline.wsj.com per cent2Farticle per cent2FSB120491426245620011.html

Brog's Product Development (2012). *Coffee Facts.* Retrieved from: www.brogscoffee.com/index.php/coffee-facts

Café de Colombia (2011). *¿Cómo enfrenta Colombia el problema de la roya?* Retrieved from: www.cafedecolombia.com/cci-fnc-es/index.php/comments/como_enfrenta_colombia_el_problema_de_la_roya/

Café de Colombia (2012). *Santander Declares A Low Coffee Leaf Rust Prevalence Zone.* Retrieved from: www.cafedecolombia.com/cci-fnc-en/index.php/comments/santander_declares_a_low_coffee_leaf_rust_prevalence_zone/

Cano, C. G., Mejía, C. V., García, E. C., Torres, J. S. A. and Calderón, E. Y. T. (2012). *El mercado mundial del café y su impacto en Colombia* (No. 009612). Banco de la República.

Caracol (2014). *En Colombia no se invierte lo suficiente en investigación sobre café*. Retrieved from www.caracol.com.co/noticias/actualidad/en-colombia-no-se-invierte-lo-suficiente-en-investigacion-sobre-cafe/20140909/nota/2406112.aspx

Carbon Disclosure Project (2014). *Reporting to CDP*. Retrieved from: www.cdp.net/en-US/Respond/Pages/companies.aspx

Carbon Disclosure Project Water (2013). *Water Program*. Retrieved from: www.cdp.net/en-US/Programmes/Pages/cdp-water-disclosure.aspx

Cárdenas, M. and Junguito, R. (2009). *Nueva Introducción a la Economía Colombiana*. Bogotá: Fundación para la Educación Superior y el Desarrollo (Fedesarrollo).

Carnegie Commission on Preventing Deadly Conflict (1997). *Preventing Deadly Conflict: Final Report*. New York: Carnegie Corporation. Retrieved from: www.ccpdc.org/pubs/rept97/finfr.htm

Carpenter, C. (2012). *Nestlé's Global Nescafé Coffee Sales Equal 4,000 Cups a Second*. 7 March. Retrieved from: www.bloomberg.com/news/2012-03-07/nestle-s-global-nescafe-coffee-sales-equal-4-000-cups-a-second.html

Castle, Timothy J. (2001). A cup fraught with issues. *Specialty Coffee Retailer* 4(4) (November).

Centro Agronómico Tropical de Investigación y Enseñanza (Tropical Agricultural Research and Higher Education Centre, CATIE) (2013). *I Regional Forum on Coffee Rust Mesoamerica. Final Report*. Turrialba, Costa Rica. Retrieved from: http://biblioteca.catie.ac.cr/royadelcafeto/descargas/Memoria_Taller_Regional_Roya.pdf

Centro de Estudios Regionales Cafeteros y Empresariales (CRECE) (2007). *Paisaje Cultural Cafetero: proceso de inclusión en la Lista de Patrimonio Mundial de Unesco, ejecución nacional, primera fase (segundo semestre de 2007), propuesta técnica y económica*. CRECE: Manizales.

Centro de Estudios Regionales Cafeteros y Empresariales (CRECE) (2011). *Monitoreo de los proyectos Nespresso en Colombia Línea de base*. CRECE: Manizales.

Centro de Estudios Regionales Cafeteros y Empresariales (CRECE) (2013). *CRECE's Monitoring & Evaluation Study on the Nespresso AAA Sustainable Quality™ Program in Colombia*. CRECE: Manizales.

Centro Nacional de Investigaciones de Café – Cenicafé (2011). Annual Report. Cenicafé.

Centro Nacional de Investigaciones de Café – Cenicafé (2012). Annual Report. Cenicafé.

Centro Nacional de Investigaciones de Café – Cenicafé (2014). *Quienes somos*. Retrieved from: www.cenicafe.org/es/index.php/quienes_somos/index.php

Chacón, M. S. (1990). *La presencia de la mujer en el desarrollo de la zona cafetera colombiana*. Bogotá: Colegio Mayor de Nuestra Señora del Rosario.

Chalarca, J. (2000). *Coffee Growing in the Department of Huila: History and Development*. Texas: University of Texas.

Chamber of Commerce of Manizales/Cámara de Comercio de Manizales (2013). *Proyectos de Infraestructura: Aerocafé*. Retrieved from: www.ccmpc.org.co/ccm/contenidos/32/AEROCAFE.pdf

Chandler, T. (2008). Nestle renegotiates Mccloud bottling contract: The Underground looks at the bigger picture. *The Trout Underground*, 5 June. Retrieved from: http://troutunderground.com/tag/predatory-multinationals/

Coffee Growers Committee of Caldas – Comité de Cafeteros de Caldas (2012). *Desarrollo de la Comunidad cafetera y su entorno. Programa Mujeres Cafeteras*. Retrieved from: http://comitedecafeteroscaldas.org/static/files/InformedelGerenteGeneral2012digital/files/assets/downloads/page0083.pdf

Coffee Research Institute (CRI) (2014). *Coffee Plant Overview: Arabica and Robusta Coffee Plant*. Retrieved from: www.coffeeresearch.org

Collier, P. and Hoeffler, A. (1998). On economic causes of civil war. *Oxford Economic Papers, 50*(4), 563–573.

Collier, P. and Hoeffler, A. (2004). Greed and grievance in civil war. *Oxford Economic Papers, 56*(4), 563–595.

Colombia Tecnología (2009). *Una Aventura en Nombre de la Nutrición*. 28 February. Retrieved from: www.colombia.com/tecnologia/autonoticias/salud/2009/02/28/DetalleNoticia 2689.asp

Colombian Agricultural Institute (ICA) (2009). *316.000 Familias Beneficiadas con Agro Ingreso Seguro*. Retrieved from: www.ica.gov.co/Noticias/Agricola-y-Pecuaria/2009/316-000-familias-beneficiadas-con-Agro-Ingreso-Seg.aspx

Committee on Sustainability Assessment (COSA) (2013). *Understanding and Promoting Sustainability through Collaboration and Science*. Retrieved from: www.theCOSA.org/

Consejo Empresarial Colombiano para el Desarrollo Sostenible (CECODES) (2012). *Sostenibilidad en Colombia. Casos Empresariales 2011*. Colombia: Consejo Empresarial Colombiano para el Desarrollo Sostenible.

Crane, A., Palazzo, G., Spence, L. J. and Matten, D. (2014). Contesting the value of the shared value concept. *California Management Review, 56*(2).

Cristancho, M. (2012). *Impacto de la Roya del Café en la Caficultura Regional*. Caldas: Centro Nacional de Investigaciones del Café (Cenicafé).

CropLife Latin America (2013). *Roya del Cafeto*. Retrieved from: www.croplifela.org/es/plaga-del-mes.html?id=29

Daneshkhu, S. (2013). Competition hots up for coffee capsule market smooth operators. *The Financial Times*, 10 November. Retrieved from www.ft.com/intl/cms/s/0/d8c237a4-489c-11e3-8237-00144feabdc0.html#axzz312btwpRs

Dauvergne, P. and Lister, J. (2012). Big brand sustainability: governance prospects and environmental limits. *Global Environmental Change, 22*(1), 36–45.

Davy, A. (2001). *Companies in Conflict Situations: A Role for Tri-Sector Partnerships?* Working Paper No. 9. Business Partners for Development.

De Ferranti, D. M. (ed.). (2005). *Beyond the City: The Rural Contribution to Development*. Washington DC: World Bank Publications.

De Ferranti, D., Perry, G., Lederman, D., Foster, W. and Valdés, A. (2005). *Más allá de la ciudad: el aporte del campo al desarrollo*. Resumen Ejecutivo. World Bank.

Departamento Administrativo Nacional de Estadística (DANE) (2013). *Comercio Internacional Exportaciones*. Retrieved from: www.dane.gov.co/index.php/comercio-y-servicios/comercio-exterior/exportaciones

Departamento de Planeación Nacional (DPN)/National Planning Department (2008). *¿En qué invierte el Estado Colombiano? Los Grandes Proyectos de Inversión del Estado Comunitario en 2008*. Agro Ingreso Seguro (AIS).

Der Grüne Punkt (2014). *Successful Management and Responsible Action*. Retrieved from: www.gruener-punkt.de

Dinero (2011). *Las Empresas más Reputadas*. 15 July. Retrieved from: www.dinero.com/edicion-impresa/caratula/articulo/las-empresas-mas-reputadas/131460

Doh, J. P. (2006). Multinational sourcing, sustainable agriculture and alleviation of global poverty. In *Multinational Corporations and Global Poverty Reduction* (pp. 235–260). Cheltenham: Edward Elgar Publications.

Domínguez, J. C. (2014). Gobierno y campesinos alcanzaron primer acuerdo en medio del paro. Minagricultura y dignidades del agro lograron cerrar el tema de refinanciación de cartera vencida. *El Tiempo*, 6 May. Retrieved from: www.eltiempo.com/archivo/documento/CMS-13939375

Dow Jones Indices (2013). *Results Announced for 2013 Dow Jones Sustainability Indices Review; 24 Sustainability Industry Group Leaders Named*. Press Release, 12 September.

Dow Jones Sustainability Indices (2013). *Industry Group Leader Report 2013*. Retrieved from: www.sustainability-indices.com

Driving Sustainable Economies (DSE) (2013). *Sector Insights: What is Driving Climate Change Action in the World's Largest Companies?* United Kingdom: Global 500 Climate Change Report 2013.

Duarte, A. (2010). *Indicadores para Valorar la Sostenibilidad de Diferentes Agrupaciones de Productores de Cafés Especiales en Colombia*. Bogotá: Universidad Nacional de Colombia.

Dube, O. and Vargas, J. F. (2006). "Resource curse in reverse: The coffee crisis and armed conflict in Colombia", Working Paper 3460, Universidad de los Andes CEDE, Bogotá.

Dube, O. and Vargas, J. F. (2007). Commodity price shocks and civil conflict: Evidence from Colombia, Working Paper 2006– 5, Royal Holloway, University of London.

Duranton, G. and Sánchez, F. (2005). *Regional Disparities in Colombia*. Documento sin publicar.

Ebrahimzadeh, C. (2012). *Dutch Disease: Too Much Wealth Managed Unwisely*. Washington, DC: International Monetary Fund.

Echandía, C. (1999). *El Conflicto Armado y las Manifestaciones de Violencia en las Regiones de Colombia*. T. I. Bogotá: Presidencia de la República, Oficina del Alto Comisionado para la Paz.

Economic Commission for Latin America and the Caribbean (ECLAC)/Comisión Económica para América Latina (CEPAL) (2002). *Centroamérica: El Impacto de la Caída de los Precios del Café en 2001*. Santiago de Chile: ECLAC.

Economic Commission for Latin America and the Caribbean (ECLAC)/Comisión Económica para América Latina (CEPAL) (2008). *Structural Change and Productivity Growth 20 Years Later: Old Problems, New Opportunities*. Santiago de Chile: CEPAL.

Economic Commission for Latin America and the Caribbean (ECLAC)/Comisión Económica para América Latina (CEPAL) and Food Agricultural Organization (FAO) (1959). *Forest Products and the Proposed Latin American Common Market*. NU. CEPAL. Session Period 8; Panama, 14–23.

Edwards, S. (1986). Commodity export prices and the real exchange rate in developing countries: coffee in Colombia. In S. Edwards and L. Ahamed (eds.), *Economic Adjustment and Exchange Rates in Developing Countries* (pp. 233–266). Chicago, IL: University of Chicago Press.

Egger, M. (2011). *Dialogue between Alliance Sud and Nestlé: The Case of Colombia*. Berne: AllianceSud.

El Nuevo Siglo (2014). *El Café y los Nuevos Mercados*. 27 August. Retrieved from www.elnuevosiglo.com.co/articulos/8-2014-el-café-y-los-nuevos-mercados.html

El País (2013). *La llegada de Starbucks Agita el Mercado Interno del Café en Colombia*. 28 August. Retrieved from: http://internacional.elpais.com/internacional/2013/08/28/actualidad/1377648132_794284.html

El Tiempo (1992). *Aporte Industrial a la Ecología*. 4 June. Retrieved from: www.eltiempo.com/archivo/documento/MAM-132135

El Tiempo (2007a). *Atentado con Carro Bomba a Planta de Nestlé en Doncello (Caquetá) Dejó Cuatro Heridos*. 18 January. Retrieved from: www.eltiempo.com/archivo/documento/CMS-3403490

El Tiempo (2007b). *Carro Bomba en Instalaciones de Nestlé Afecta Principalmente a 1.400 Empleados.* 19 January. Retrieved from: www.eltiempo.com/archivo/documento/CMS-3404721

El Tiempo (2007c). *Ludoteca para Enseñar Derechos de la Niñez en Bugalagrande (Valle).* 21 September. Retrieved from: www.eltiempo.com/archivo/documento/CMS-3733982

El Tiempo (2008). *En Dosquebradas, Maloka y Nestlé se Unen por la Nutrición, la Salud y el Bienestar de los Jóvenes.* 16 August. Retrieved from: www.eltiempo.com/archivo/documento/CMS-4451545

El Tiempo (2010). *Café 'Variedad Colombia', Solución contra la Roya.* 8 October. Retrieved from: www.eltiempo.com/colombia/eje-cafetero/articulo-web-new_nota_interior-8095302.html

El Tiempo (2011). *Las FARC ¿Tras el Control de la Industria Lechera en el Caquetá?* 26 February. Retrieved from: www.eltiempo.com/archivo/documento/MAM-2399388

El Tiempo (2014). *MinInterior reanuda el diálogo con sectores en paro. Habrá reunión esta tarde en la que se esperan acuerdos.* 8 May. Retrieved from: www.eltiempo.com/politica/gobierno/mininterior-reanuda-el-dialogo-con-sectores-en-paro/13959077

Emprender Paz (n.d.). *Banco de Buenas Prácticas: Sector Privado y Construcción de Paz* [Initiative]. Retrieved from: www.emprenderpaz.org/descargas/ARGOS_Buenaspracticas.pdf

Environmental Leader (2012). *Nestlé Nespresso Hits 75 per cent Capsule Recovery Goal.* 12 September. Retrieved from: www.environmentalleader.com/2012/09/12/nestle-nespresso-hits-75-capsule-recycling-goal/

Ernst Basler + Partner (n.d.). *Energetic Use of Residues from Coffee Production in Central and South America.* Renewable Energy & Energy Efficiency Promotion in International Cooperation (REPIC), Swiss Confederation. Retrieved from: www.ebp.ch/files/projekte/rk_kaffeeabfaelle_11_introduction_coffee_steiner.pdf

Ethical Coffee (n.d.). *Rainforest Alliance Certification.* Retrieved from: www.ethicalcoffee.net/rainforest.html

Ethical Consumer (2013). *About Us.* Retrieved from: www.ethicalconsumer.org

Ethical Corporation (2014). *Ethical Corporation's Responsible Business Awards 2014.* Retrieved from: http://events.ethicalcorp.com/awards/past-winners.php

Evans, A., Cabral, L. and Vadnjal, D. (2006). *Sector-Wide Approaches in Agriculture and Rural Development, Phase I: A Desk Review of Experience, Issues and Challenges.* Bonn, Germany: Global Donor Platform for Rural Development.

Fairtrade International (FLO) (2009). *A Charter of Fair Trade Principles.* The Netherlands: World Fair Trade Organization and Fairtrade Labelling Organizations International.

Fairtrade International (FLO) (2011). *Coffee.* Retrieved from: www.fairtrade.net/coffee.html

Fairtrade International (FLO) (2013). *Nespresso and Fairtrade Join Forces* [Launch Innovative Farmer Support Programme]. 16 July. Retrieved from: www.fairtrade.net/single-view+M5a5d82d005a.html.

Falck, O. and Heblich, S. (2007). Corporate social responsibility: doing well by doing good. *Business Horizons, 50*(3), 247–254.

FAO (2013). *Alianzas público-privadas para el desarrollo de agronegocios – Informe de país: Colombia.* Estudios de casos de países – América Latina. Rome.

Federación Nacional de Cafeteros (FNC) (2007). *Renovación Compromiso Cafetero.* Annual Reports to the Coffee Congress. Bogota: FNC.

Federación Nacional de Cafeteros (FNC) (2008). *De Acuerdo con la Prosperidad. Annual Reports to the Coffee Congress.* Bogota: FNC.

Federación Nacional de Cafeteros (FNC) (2010). *Prosperidad Cafetera: Informe del Gerente General al LXXV Congreso Nacional de Cafeteros Bogotá.* Bogotá: FNC.

Federación Nacional de Cafeteros (FNC) (2011a). *Comportamiento de la Industria Cafetera.* Bogotá: FNC

Federación Nacional de Cafeteros (FNC) (2011b). Variables used to determine the domestic reference price of Colombian coffee. *Colombian Insider: The Coffee World.* Retrieved from: www.federaciondecafeteros.org/algrano-fnc-en/index.php/comments/variables_used_to_determine_the_domestic_reference_price_for_colombian_coff

Federación Nacional de Cafeteros de Colombia (FNC) (2012a). *FNC is Recognized in Rio+20 as One of the Top Companies in Productive Transformation.* Bogotá: FNC.

Federación Nacional de Cafeteros de Colombia (FNC) (2012b). *Sostenibilidad en Acción* (Report). Bogotá: FNC.

Federación Nacional de Cafeteros de Colombia (FNC) (2012c). *Colombia: World Leader in Value Added and Sustainable Coffee Marketing.* Retrieved from: www.federaciondecafeteros.org/particulares/en/sala_de_prensa/detalle/colombia_world_leader_in_value_added_and_sustainable_coffee_marketing/

Federación Nacional de Cafeteros de Colombia (FNC) (2012d). *Mujeres cafeteras de Colombia ganan terreno.* Retrieved from: www.federaciondecafeteros.org/algrano-fnc-es/index.php/comments/mujeres_cafeteras_de_colombia_ganan_terreno

Federación Nacional de Cafeteros de Colombia (FNC) (2013a). *Colombia Sigue Demostrando Importantes Avances en Productividad Cafetera.* Bogotá: FNC. Retrieved from: www.federaciondecafeteros.org/particulares/es/sala_de_prensa/detalle/colombia_sigue_demostrando_importantes_avances_en_productividad_cafetera/

Federación Nacional de Cafeteros de Colombia (FNC) (2013b). *FNC en Cifras.* Bogotá: FNC. Retrieved from: www.federaciondecafeteros.a/particulares/es/quienes_somos/fnc_en_cifras/Nespresso

Federación Nacional de Cafeteros de Colombia (FNC) (2013c). *La Producción Registrada de Julio 2013 Supero un Millón de Sacos.* Bogotá: FNC.

Federación Nacional de Cafeteros de Colombia (FNC) (2013d). *Nespresso S.A Announces New Specialty Coffee Capsules from Two Colombian Region, Cauca and Santander.* Retrieved from: www.federaciondecafeteros.org/particulares/en/sala_de_prensa/detalle/nespresso_s.a_announces_new_specialty_coffee_capsules_from_two_colombian_re/

Federación Nacional de Cafeteros de Colombia (FNC) (2013e). *Cafeteros de Colombia han Recibido más de \$806.000 Millones bajo los Programas AIC-PIC.* 22 October. Retrieved from: www.federaciondecafeteros.org/particulares/es/sala_de_prensa/detalle/cafeteros_de_colombia_han_recibido_mas_de_806.000_millones_bajo_los_program/

Federación Nacional de Cafeteros de Colombia (FNC) (2014a). *Colombian Coffee Growers and Nespresso Commemorate 10 Year.* Retrieved from: www.federaciondecafeteros.org/particulares/en/sala_de_prensa/detalle/colombian_coffee_growers_and_nespresso_commemorate10_year/

Federación Nacional de Cafeteros de Colombia (FNC) (2014b). *Price Upswing Brings Relief and Moderate Optimism to Coffee Growers.* Retrieved from: www.federaciondecafeteros.org/algrano-fnc-en/index.php/comments/price_upswing_brings_relief_and_moderate_optimism_to_coffee_growers/

Federación Nacional de Cafeteros (FNC) and Nespresso (2013). *Team Colombian Coffee Growers Federation and Nestlé Nespresso SA.* Bogotá: FNC.

Fieser, E. (2009). Fair Trade: what price for good coffee? Fair Trade practices were created to help small farmers, but they may have hit their limits. *Time,* 5 October. Retrieved from: http://content.time.com/time/magazine/article/0,9171,1926007,00.html

Flórez, A. (2010). *Perspectivas del Mercado de Cafés Sostenibles*. Seminario Internacional sobre Café Sostenible, Bogotá, Colombia.

Flury, K. (2012). "Rabobank Agri Commodity Markets Research. Market Outlook – Era of Oversupply." Rabobank International, London. Retrieved from: www.sintercafe. com/uploads/File/2012/presentations/10.flury.pdf

Forum del Café (2009). *Los Cafés de Fincas*. Forum Cultural del Café. Retrieved from: www.forumdelcafe.com

Friedman, M. (1970). The social responsibility of business is to increase its profits. *The New York Times Magazine,* 13 September.

FTSE International Limited (2006). *FTSE4Good Index Series Inclusion Criteria.* Retrieved from: www.ftse.com/Indices/FTSE4Good_Index_Series/Downloads/FTSE4Good_Inclusion_Criteria.pdf

Fundación Felipe Gómez Escobar (2014). *Nuestras Alianzas*. Retrieved from: https://juanfe. org

Fundación Ideas para la Paz (2011). *Caso Centro de Reconciliación Nestlé-Bugalagrande*. Retrieved from: www.ideaspaz.org/portal/images/Caso per cent20Centro per cent20de per cent20Reconciliacion per cent20Bugalagrande per cent20- per cent20Nestle.pdf

Fundación para la Educación Superior y el Desarrollo (Fedesarrollo), Centro de Investigación Económica y Social (2012). *Consultoría sobre Costos de Producción de Doce Productos Agropecuarios*. Bogota: Fedesarrollo.

Fundación Sarmiento Palau-Nestlé de Colombia (2009). *Informe de Actividades Mayo 1 de 1999-Abril 30 de 2000*. Programa de Desarrollo Empresarial, Bugalagrande, Colombia.

Furrer, O. (2011). A customer relationship typology of product services strategies. In F. Gallouj and F. Djellal (eds). *The Handbook of Innovation and Services: A Multi-disciplinary Perspective*. Cheltenham: Edward Elgar Publishing.

Galindo, H., Restrepo, J. and Sánchez, F. (2009). Conflicto y Pobreza en Colombia: Un Enfoque Institucionalista. In J. Restrepo and D. Aponte (eds), *Guerra y Violencias en Colombia Herramientas e Interpretaciones* (pp. 315–351). Bogotá: Pontificia Universidad Javeriana.

Garzón, C. M. (2002). *Mujeres Trabajadoras del Café*. Bogotá: Ministerio de Cultura.

Genesis Foundation (n.d.). *Nutrir.* Retrieved from: www.genesis-foundation.org/fundacin-nutrir/

Giovannucci, D. and Koekoek, F. J. (2003). The state of sustainable coffee: A study of twelve major markets. International Coffee Organization, International Institute for Sustainable Development and UNCTAD.

Giovannucci, D. and Ponte, S. (2005). Standards as a new form of social contract? Sustainability initiatives in the coffee industry. *Food Policy, 30,* 284–301.

Giovannucci, D. and Purcell, T. (2008c). *Standards and Agricultural Trade in Asia*. Institute Discussion Paper No. 107. Mandaluyong City, Philippines: Asian Development Bank (ADB).

Giovannucci, D. and Samper, L. (2009). The Case of Café Nariño, Colombia. In D. Giovannucci, T. Josling, W. Kerr, B. O'Connor and M. Yeung (eds), *Guide to Geographical Indications: Linking Products and their Origins* (pp. 197–202). Geneva: International Trade Centre.

Giovannucci, D., Leibovich, J., Pizano, D., Paredes, G., Montenegro, S., Arévalo, H. and Varangis, P. (2002). *Colombia Coffee Sector Study* (Report No. 24600-CO). Washington, DC: World Bank.

Global Alliance for Improved Nutrition (2013). *Access to Nutrition Index: Nestlé SA*. Retrieved from: www.accesstonutrition.org/nestlé-sa

Global Envision (2005). *Wake up, Smell the Coffee.* 26 April. Retrieved from: www.globalenvision.org/library/6/743

Global Exchange (2011). *Coffee FAQ.* Retrieved from: www.globalexchange.org/fairtrade/coffee/faq

Gobierno de Colombia (2013). *Colombia Estadísticas: Salario Mínimo Legal.* Retrieved from: www.colombia.com/colombiainfo/estadisticas/salario.asp

González, C. (2010). *Programa de Cafés Sostenibles de la Federación Nacional de Cafeteros.* Memorias del Seminario Internacional sobre Café Sostenible, Bogotá, Colombia.

Government of Antioquia (n.d.). *Food Security and Nutritional Plan of Antioquia, MANA, Republic of Colombia* (Mejoramiento Alimenticio y Nutricional de Antioquia). Retrieved from: http://unpan1.un.org/intradoc/groups/public/documents/un-dpadm/unpan046051.pdf

Government of Valle del Cauca (2007). *Se inauguró Ludoteca en Bugalagrande.* 19 September. Retrieved from: www.valledelcauca.gov.co/publicaciones.php?id=3246&dPrint=1

Granados, J. (2010). *Las Migraciones Internas y su Relación con el Desarrollo en Colombia: Una Aproximación desde Algunos Estudios no Clasificados como Migración Interna de los Últimos 30 Años.* Bogotá: Pontificia Universidad Javeriana.

Griffin, K. (1968). Coffee and the economic development of Colombia. *Bulletin of the Oxford University Institute of Economics & Statistics, 30*(2), 105–127.

Grupo Éxito (2012). *Winners 'Proveedores de Éxito Award'.* Colombia. Retrieved from: www.grupoexito.com.co/index.php/en/component/content/article/27proveedores/proveedores-de-exito/756-winners-proveedores-de-exito-2012

Grupo Éxito (2013). *Suppliers FAQ.* Retrieved from: www.grupoexito.com.co/index.php/en/suppliers/proveedores-de-exito-award/faqs

Guarín, S., Navarro, H. and Pellerano, L. (2008). *Evaluación del Impacto de los Programas de Paz y Desarrollo y Laboratorios de Paz: Línea de Base e Impactos Preliminares.* Bogotá: Departamento Nacional de Planeación.

Guthrie, D. (2014). *A Conversation on Corporate Social Responsibility.* Forbes, 9 January. Retrieved from: www.forbes.com/sites/dougguthrie/2014/01/09/a-conversation-on-corporate-social-responsibility/

Hall, R. E. and Jones, C. I. (1999). Why do some countries produce so much more output per worker than others? *The Quarterly Journal of Economics, 114*(1), 83–116.

Haynes-Peterson, R. (2014). *Nespresso Launches Limited Edition Colombian Terroirs Coffees for Spring.* Retrieved from: www.examiner.com/article/nespresso-launches-limited-edition-colombian-terroirs-coffees-for-spring

Hebebrand, C. (2011). Leveraging private sector investment in developing country agrifood systems. *Policy Paper Series. The Chicago Council on Global Affairs.*

Henríquez, J. L. (2003) El café perdió el año, *El Diario de Hoy* (El Salvador) 11 October.

Heynike, P. (2010). *Nestlé: The No 1. Beverage Powerhouse.* Presented at the Nestlé Investor Seminar, 21 June, Montreux, Switzerland.

Hillemanns, C. F. (2003). UN Norms on the responsibilities of transnational corporations and other business enterprises with regard to human rights. *European & International Law, 4*(10), 1065–1080.

Hills, G., Russell, P., Borgonovi, V., Doty, A. and Iyer, L. (2012) *Shared Value in Emerging Markets. How Multinational Corporations Are Redefining Business Strategies to Reach Poor or Vulnerable Populations.* Boston, MA: Foundation Strategy Group (FSG).

History of Business (2008). *History of Nescafé.* Retrieved from: http://historyofbusiness.blogspot.mx/2008/09/history-of-nescafe.html

Humphrey, J. (2006). *Global Value Chain in the Agrifood Sector*. Vienna: United Nations Industrial Development Organization (UNIDO).

Hwy-Chang, M., Parc, J., So Hyun, Y. and Nari, P. (2011). An extension of Porter and Kramer's creating shared value (CSV): reorienting strategies and seeking international cooperation. *Journal of International & Area Studies, 18*(2), 49–64.

Industria Alimenticia (2009). *El Premio Nestlé en Creación de Valor Compartido*. 7 October. Retrieved from: www.industriaalimenticia.com/articles/el-premio-nestle-en-creacion-de-valor-compartido

Inter-American Development Bank (2013a). *Creating Shared Value: The Changing Role of Business in Development*. Retrieved from: http://events.iadb.org/calendar/eventDetail. aspx?lang=En&id=4221

Inter-American Development Bank (2013b). *Future Finance: Private Sector with Purpose*. Retrieved from: www.iadb.org/en/structured-and-corporate-finance/future-finance-private-sector-with-purpose,6932.html

Institute for Development Studies (IDS) (2014). *Research and Practical Analysis on the Impacts of Business on Development and the Role of Governments, Donors and NGOs*. Retrieved from: www.ids.ac.uk/files/dmfile/IDSBusinessflyer.pdf

Instituto Centroamericano de Administración de Empresas (INCAE)/Central American Institute of Business Administration (2013). *CIMS, INCAE and Nespresso Presented First MBA Challenge Prize*. Retrieved from: www.incae.edu/es/en/news/cims-incae-and-nespresso-presented-first-mba-challenge-prize.php

Instituto Colombiano Agropecuario (2009). Resolución derogada por el artículo 12 de la Resolución 1183 de 2010. *Diario Oficial No. 47.483*, 25 September.

Instituto Geográfico Agustín Codazzi (IGAC) (2013). *Nuestra identidad*. Bogotá, Colombia. Retrieved from: www.igac.gov.co/igac

International Coffee Organization (ICO) (2014). *Sustainability Initiatives*. Retrieved from: www.ico.org/sustaininit.asp?section=About_Coffee

International Trade Centre (ITC) (2010). *Climate Change and the Coffee Industry*. Geneva: ITC.

International Trade Centre (ITC) (2011). *The Coffee Exporter's Guide*. Geneva: ITC.

Jaffee, D. (2012). Weak coffee: Certification and co-optation in the fair trade movement. *Social Problems, 59*(1), 94–116.

Jolly, D. (2014). Nestlé loses a clash over single-serve coffee. *The New York Times*, 17 April. Retrieved from: www.nytimes.com/2014/04/18/business/international/nestle-loses-a-clash-over-single-serve-coffee.html?_r=0

Josephs, L. (2014). Dry weather in Brazil boosts sugar, coffee prices. *The Wall Street Journal*, 24 February. Retrieved from: http://online.wsj.com/news/articles/SB10001424052702 3034263045790435422077767638

Juan Valdez (2014). *Sostenible desde el Origen*. Retrieved from: www.juanvaldezcafe.com

Junguito, R. and Pizano, D. (1991). Producción de café en Colombia. *FEDESARROLLO: Fondo Cultural Cafetero*.

Jurado, C. and Tobasura, I. (2012). Dilema de la Juventud en Territorios Rurales de Colombia: ¿Campo o Ciudad? *Revista Latinoamericana de Ciencias Sociales, Niñez y Juventud, 10*(1), 63–77.

Kennedy, E. H., Beckley, T. M., McFarlane, B. L. and Nadeau, S. (2009). Why we don't "walk the talk": Understanding the environmental values/behaviour gap in Canada. *Human Ecology Review, 16*(2): 151–160.

Kolk, A. (2012). Towards a sustainable coffee market: Paradoxes faced by a multinational company. *Corporate Social Responsibility and Environmental Management, 19*(2), 79–89.

Kytle, B. and Ruggie, J. G. (2005). *Corporate Social Responsibility as Risk Management: A Model for Multinationals.* Corporate Social Responsibility Initiative Working Paper No. 10. Cambridge, MA: John F. Kennedy School of Government.

La Tarde (2014). *Nescafé Plan llegó para apoyar a los cafeteros y asegurar un nuevo Mercado.* 31 January. Retrieved from: www.latarde.com/noticias/economica/128753-nescafe-plan-llego-para-apoyar-a-los-cafeteros-y-asegurar-un-nuevo-mercado

Lantos, G. P. (2001). The boundaries of strategic corporate social responsibility. *Journal of Consumer Marketing, 18*(7), 595–632.

Las Páginas Amarillas de Florencia (2014). *Caquetá.* [Información del Municipio de Florencia]. Retrieved from: www.laspaginasamarillasdecolombia.com/caqueta/florencia. html

Latham, J. (2012). *Way Beyond Greenwashing: Have Corporations Captured Big Conservation?* Retrieved from: www.independentsciencenews.org/environment/way-beyond-greenwashing-have-multinationals-captured-big-conservation/

Leibovich, J. and Botello, S. (2008). Análisis de los cambios demográficos en los municipios cafeteros y su relación con los cambios en la caficultura Colombiana (1993–2005). *Ensayos sobre Economía Cafetera,* No. 24.

Leibovich, J. and García, A. (2010). *Proyecto COSA: Línea de Base. Evaluación de Impacto de la Adopción de Prácticas Sostenibles en la Producción de Café en Colombia.* Manizales: Centro de Estudios Regionales Cafeteros y Empresariales (CRECE).

Leiteritz, R., Nasi, C. and Rettberg, A. (2009). Para desvincular los recursos naturales del conflicto armado en Colombia. Recomendaciones para Formuladores de Política y Activistas. *Colombia Internacional, 70,* 215–229.

Lewis, W. A. (1954). Economic development with unlimited supply of labour. *The Manchester School, 22*(2), 139–191.

Linton, A. (2005). Partnering for sustainability: business–NGO alliances in the coffee industry. *Development in Practice, 15*(3–4), 600–614.

Lisboa-Bacha, E. et al. (1993). *150 Años de Café.* Federación Nacional de Cafeteros de Colombia (FNC). Gerencia Comercial, Inteligencia Competitiva.

Lozano, A. (2007). Relaciones de tamaño, producción y trabajo en las fincas cafeteras colombianas. *Ensayos sobre Economía Cafetera, 22,* 85–106.

Lozano, A. (2009). Acceso al crédito en el sector cafetero Colombiano. *Ensayos sobre Economía Cafetera, 25,* 95–121.

Lozano, M. M. (2010). La Caficultura Colombiana en el Siglo XXI: Una Revisión de la Literatura Reciente. *Revista Gestión y Región* (9).

Mandell, E. (2013). Helping coffee farmers turns competitors into colleagues. *Global Envision,* 3 June. Retrieved from: www.globalenvision.org/2013/06/03/interview-helping-coffee-farmers-turns-competitors-colleagues

Manning, S., Boons, F., Von Hagen, O. and Reinecke, J. (2012). National contexts matter: The co-evolution of sustainability standards in global value chains. *Ecological Economics, 83,* 197–209.

Market Wired (2013). *The Colombian Coffee Growers Federation Exports Record Number of Specialty Coffee Bags.* 21 October. Retrieved from: www.marketwired.com/press-release/colombian-coffee-growers-federation-exports-record-number-specialty-coffee-bags-1843122.htm

Matzler, K., Bailom, F., von den Eichen, S. and Kohler, T. (2013). Business model innovation: coffee triumphs for Nespresso. *Journal of Business Strategy, 34*(2), 30–37.

McMillan, M. and Rodrik, D. (2011). *Globalization, Structural Change, and Productivity Growth.* Geneva: International Labour Organization and World Trade Organization.

Mejía, A. C. (2012). *Colombia: Coffee Production and Institutional Framework.* Bogotá: Federación Nacional de Cafeteros.

Mejía, O. W. (2011). *Diseño Y Formulación Operativa de una Política Integral de Gestión Migratoria Laboral y las Herramientas que Permitan su Desarrollo* (Informe Final No. 1351). Pereira: Centro de Investigación y Desarrollo Universidad Nacional de Colombia (CID).

Mejía Cubillos, J. (2012). *Agro, Ingreso Seguro in a Public Policy Analysis Perspective.* MPRA Paper No. 39998. Retrieved from: http://mpra.ub.uni-muenchen.de/39998/

Mendez, V. E., Bacon, C. M., Olson, M., Petchers, S., Herrador, D., Carranza, C., ... and Mendoza, A. (2010). Effects of Fair Trade and organic certifications on small-scale coffee farmer households in Central America and Mexico. *Renewable Agriculture and Food Systems, 25*(3), 236–251.

Ministerio de Agricultura y Desarrollo Rural (2005). *La Cadena del Café en Colombia. Una Mirada Global de su Estructura y Dinámica, 1991–2005* (Documento de Trabajo No. 59). Bogotá: Observatorio Agrocadenas.

Ministerio de Cultura (2011). *Paisaje Cultural Cafetero: Un Paisaje Cultural Productivo en Permanente Desarrollo.* Taller Editorial Escuela Taller de Bogotá, Colombia.

Ministerio del Trabajo (2013). "Salario Mínimo" Retrieved from: www.mintrabajo.gov. co/empleo/abece-del-salario-minimo.html

Ministerio del Trabajo (2014). *Caficultores de Caldas tendrán Protección en su Vejez.* Retrieved from: www.mintrabajo.gov.co/marzo-2014/3126.html

Ministry of Agriculture and Rural Development and National Planning Department (2011). *National Evaluation System.* Levantamiento de Información y Evaluación de los Resultados de la Ejecución del Programa Agro Ingreso Seguro (AIS). Retrieved from: https:// sinergia.dnp.gov.co/Sinergia/Archivos/5738ccc4-1da2-47ec-a59a-c9c2df8682ce/ INFORME_FINAL_Agro_Ingreso_Seguro.pdf

Ministry of Agriculture and Rural Development (2013). *Programa Desarrollo Rural con Equidad (DRE).* Retrieved from: www.minagricultura.gov.co/ministerio/programas-y-proyectos/Paginas/Programa-Desarrollo-Rural-con-Equidad-DRE.aspx

Monson, V. (1991). The story of Nescafe. *Nutrition & Food Science, 91*(1), 8–9.

Moore, G. (2004). The Fair Trade movement: parameters, issues and future research. *Journal of Business Ethics, 53*, 73–86.

Mora, T. and Tovar, D. (2011) *Declaración de la CUT Colombia.* Central Unitaria de Trabajadores de Colombia.

Muñoz, L. G. (2010). Editorial: Acuerdo por la Prosperidad Cafetera 2010–2015. *Ensayos sobre Economía Cafetera,* No. 26. Federación Nacional de Cafeteros de Colombia.

Muñoz, L. G. (2013). *Discurso Gerente General de la Federación de Cafeteros.* Speech presented at the LXXVII Assembly of Asoexport, 13 November. Cartagena, Colombia.

Muradian, R. and Pelupessy, W. (2005). Governing the coffee chain: the role of voluntary regulatory systems. *World Development, 33*(12), 2029–2044.

Nates Cruz, B. and Velásquez López, P. (2011). Territorios en mutación Crisis cafetera, crisis del café. *Cuadernos de Desarrollo Rural, 6*(63), 22.

National Coffee Park (n.d.). *Historia.* Retrieved from: www.parquenacionaldelcafe.com/ newpage/culturacafetera.php

Neary, H. (1997) Equilibrium structure in an economic model of conflict. *Economic Inquiry, 35*(3), 480–494.

Neilson, J. and Pritchard, B. (2007). Green coffee? The contradictions of global sustainability initiatives from an Indian perspective. *Development Policy Review, 25*(3), 311–331.

Neilson, J. (2008). Global private regulation and value-chain restructuring in Indonesian smallholder coffee systems. *World Development, 36*(9), 1607–1622.

Nelson, J. (2000). *Business of Peace. The Private Sector as a Partner in Conflict Prevention and Resolution.* International Business Leaders Forum (IBLF). Birmingham: Folium Press.

Nescafé (n.d.). *Responsibility Goes Beyond the Cup: The Nescafé Plan: Creating Share Value.* Retrieved from: www.nescafe.com/sustainability

Nescafé (2013). *All You Want to Know about Coffee. Coffee History.* Retrieved from: www.nescafe.com/coffee_history_en_com.axcms

Nescafé Dolce Gusto (2013). *About Nescafé Dolce Gusto.* Retrieved from: www.dolce-gusto.us

Nescafé Mexico (2014). *Responsible Farming.* Retrieved from: www.nescafe.com.mx/farming02_en_co_uk.axcms

Nestlé Colombia (2006). *Medio Ambiente y Desarrollo Sostenible.* Nestlé SA. Zetta Comunicadores SA.

Nestlé Colombia (2013a). *Nestlé crea valor compartido en Colombia Informe de creación de valor compartido 2012.*

Nestlé Colombia (2013b). *Plan Nescafé®.* Retrieved from: http://corporativa.nestle.com.co/csv/desarrollo-rural/plan-nescafÉ

Nestlé Colombia (2014). *Nestlé renueva los cafetales de Risaralda con expansión del Plan Nescafé.* Retrieved from: http://corporativa.nestle.com.co/media/pressreleases/nestle-renueva-los-cafetales-de-risaralda

Nestlé Nespresso SA (2008). *Nespresso AAA Sustainable Quality Program: Creating Value, Sharing Value.* The Second Nestlé Nespresso AAA Sustainable Quality Coffee Forum.

Nestlé Nespresso SA (2009a). *Media Brief 2.* Nespresso AAA Sustainable Quality Program.

Nestlé Nespresso SA (2009b). *Herramienta TASQ Genérica Versión 1009.*

Nestlé Nespresso SA (2009c). *Herramienta TASQ Módulo Colombia. Versión 1009.*

Nestlé Nespresso SA (2010). *Leidenschaftliches Streben nach Höchstleistungen.*

Nestlé Nespresso SA (2011a). *Nespresso Ecolaboration™ Mid-Term Report.* Nestlé Nespresso SA, Corporate Communications.

Nestlé Nespresso SA (2011b). *Real Farmer Income Cases Studies.*

Nestlé Nespresso SA (2011c). *Ensuring Abundant Supply of High Quality Coffee.*

Nestlé Nespresso SA (2012a). *AAA Sustainable Quality™ MBA Challenge 2013.* Background on Nestlé Nespresso SA Sustainable Markets Intelligence Center (CIMS) and INCAE Business School. Retrieved from: http://sustainabilitymbachallenge.com/wp-content/uploads/2012/09/MBA_Challenge.pdf

Nestlé Nespresso SA (2012b). *TASQ™ Auto-Evaluación. Herramienta para la Evaluación de la Calidad Sostenible: Colombia.*

Nestlé Nespresso SA (2013a). *Students of Rotterdam School of Management (Erasmus University) Get to the Finals of the Sustainable.* Retrieved from: www.careersatnespresso.com/assets/Nespresso_MBA_Challenge-EN.pdf

Nestlé Nespresso SA (2013b). *Accelerating Progress on the Nespresso AAA Sustainable Quality™ Program in Central America.*

Nestlé Nespresso SA (2013c). *Team Colombian Coffee Growers Federation and Nestlé Nespresso.*

Nestlé Nespresso SA (2013d). *Helping Farmers to Share their Workload in Colombia.*

Nestlé Nespresso SA (2013e). *Helping Farmers Improve their Business Skills in Huehuetenango, Guatemala.*

Nestlé Nespresso SA (2013f). *Improving our Environmental Impact from the Cherry to the Cup.*

Nestlé Nespresso SA (2013g). *Perfecting our Packaging Solution.*

Nestlé Nespresso SA (2014a). *The Grand Crus.* Retrieved from: www.nespresso.com/#/cl/en/coffee_nespresso/grands_crus

Nestlé Nespresso SA (2014b). *Nespresso Colombian Terroirs Limited Edition.* Retrieved from: www.nespresso.com/#/cl/en/coffee_nespresso/grands_crus

Nestlé Nespresso SA (2014c). *Capsule Recycling*. Retrieved from: www.nestle-nespresso. com/ecolaboration/sustainability/capsules

Nestlé Nespresso SA (2014d). *Nespresso to Enter Colombian Market*. Retrieved from: www. nestle.com/media/newsandfeatures/nespresso-enters-colombian-market

Nestlé Nespresso SA (2014e). *Nespresso Announces Opening of Colombian Market*. Retrieved from: www.nestle-nespresso.com/newsandfeatures/nespresso-announces-opening-of-colombian-market

Nestlé Nespresso SA (2014f). *Background on Nestlé Nespresso SA Nespresso Sustainability MBA Challenge 2014. Pushing the Boundaries of Sustainable Quality Coffee Sourcing*. Retrieved from: http://sustainabilitymbachallenge.com/wpcontent/uploads/2012/09/ Backgrounder -2014.pdf

Nestlé Nespresso SA (2014g). *Nespresso to Invest CHF 500 Million into 2020 Sustainability Strategy*. Retrieved from: www.nestle-nespresso.com/newsandfeatures/nespresso-to-invest-chf-500-million-into-2020-sustainability-strategy

Nestlé Nespresso SA (2014h). *Protecting, Regenerating and Improving the Coffee Ecosystems and Communities where Nespresso is Involved through an Extensive Agro-forestry Program*.

Nestlé Nespresso SA (2014i). *Nespresso Launches its 2020 Sustainability Ambition, The Positive Cup, and the Nespresso Sustainable Development Fund*. Retrieved from: www.nestle-nespresso.com/media/mediareleases/nespresso-launches-its-2020-sustainability-ambition-the-positive-cup-and-the-nespresso-sustainable-development-fund?view=more

Nestlé SA (2003). *Informe sobre Desarrollo Humano de Nestlé*. Nestlé SA, Public Affairs, Zurich.

Nestlé SA (2004). *Las Caras del Café* (Reporte de Nestlé sobre el Café). Nestlé SA, Public Affairs, Lausanne.

Nestlé SA (2005). *El Concepto de Responsabilidad Social Corporativa de Nestlé Según se ha Implementado en Latinoamérica*. Nestlé SA, Public Affairs, Lausanne.

Nestlé SA (2006). *Nestlé, la Comunidad y los Objetivos de Desarrollo del Milenio de las Naciones Unidas*. Nestlé SA, Public Affairs, Lausanne.

Nestlé SA (2008). *Informe sobre Creación de Valor Compartido*. Nestlé SA, Public Affairs, Lausanne.

Nestlé SA (2009). *Nespresso and Colombian Coffee Growers Federation Announce New 'Pure Origin' Grand CRU Coffee Composed Entirely of Colombian Arabicas*. Retrieved from: http://corporativa.nestle.com.co/media/pressreleases/2009_fnceventrosabaya

Nestlé SA (2010). *Nestlé de Colombia y la Alcaldía Municipal Inauguran Centro de Reconciliación en Bugalagrande, Valle del Cauca*. Retrieved from: http://corporativa.nestle.com.co/ media/pressreleases/2010_centroreconciliacionbugalagrande

Nestlé SA (2011). *Nestlé Investor Seminar*. Retrieved from: www.nestle.com/asset-library/ documents/library/presentations/investors_events/investor_seminar_2011/nis2011-02-supply-chain-competitive-gaps-hjoehr.pdf

Nestlé SA (2012a). *Energy from Coffee Grounds, Colombia*. Retrieved from: www.nestle.com/ csv/case-studies/allcasestudies/energy-spent-coffee-grounds-colombia

Nestlé SA (2012b). *The Nescafé Plan*. Retrieved from: www.nestle.com/csv/case-studies/ AllCaseStudies/TheNescaféPlan-Mexico

Nestlé SA (2012c). *New Nespresso Limited Edition Coffee Draws on Wine-Making Technique*. Retrieved from: www.nestle.com/media/newsandfeatures/nespresso-wine-making

Nestlé SA (2013a). *Nestlé among Top Performers in New Access to Nutrition Index*. Retrieved from: www.nestle.com/media/newsandfeatures/atni

Nestle SA (2013b). *Nestlé CDP Water Assessment 2013*. Retrieved from: www.nestle.com/ asset-library/documents/creating per cent20shared per cent20value/performance/ nestle-response-to-cdp-2013-water.pdf

Nestlé SA (2013c). *Nespresso-Recycling@Home*. Retrieved from: www.nestle.com/csv/case-studies/AllCaseStudies/Nespresso_Recycling

Nestlé SA (2013d). *Nespresso Launches Major New Sustainability Initiatives in Africa and Latin America*. Retrieved from: www.nestle.com/media/newsandfeatures/nespresso-sustainability-advisory-board

Nestlé SA (2014a). *New Pilot Retirement Fund for Colombian Coffee Farmers*. Retrieved from: www.nestle.com/media/news/nespresso-pensions-colombian-coffee-farmers

Nestlé SA (2014b). *Creating Shared Value and Meeting our Commitments 2013*.

Nestlé SA (2014c). *Nescafé Launches Shakissimo, a Chilled Coffee Range for Europe*. Retrieved from: www.nestle.com/media/news/nestle-launches-shakissimo

Nestlé SA (2014d). *World's Favourite Coffee Brand, Nescafé, Launches REDvolution*. Retrieved from: www.nestle.com/Media/NewsAndFeatures/Nescafe-REDvolution

Nestlé UK (2014). *First Harvest Celebrations for the Nescafé Plan in Colombia*. Retrieved from: www.nestle.co.uk/media/pressreleases/nestle-first-harvest.

Ocampo, J. (1990). Import controls, prices and economic activity in Colombia. *Journal of Development Economics*, *32*(2), 369–387.

Organisation for Economic Co-operation and Development (OECD) (2002). *Multinational Enterprises in Situations of Violent Conflict and Widespread Human Rights Abuses*. Directorate for Financial, Fiscal and Enterprise Affairs Working Papers on International Investment, No. 2002/1.

Organisation for Economic Co-operation and Development (OECD) (2008). *Latin American Economic Outlook 2009*. Paris: OECD.

Organisation for Economic Co-operation and Development (OECD) (2010). *Colombian Economic Assessment*. Paris: OECD.

Ornelas, A. (2007). *Nestlé Enfrenta Boicot de las FARC en Colombia*. Swiss Broadcasting Corporation. Retrieved from: www.swissinfo.ch/spa/actualidad/Nestle_enfrenta_boicot_de_las_FARC_en_Colombia.html per cent3Fcid per cent3D5822292&client=safari&hl=en&gl=mx&strip=1

Oxfam International (2014). *Behind the Brands*. Retrieved from: www.oxfam.org/en/grow/campaigns/behind-brands

Palacios, M. (1980). *Coffee in Colombia, 1850–1970: An Economic, Social, and Political History*. Cambridge: Cambridge University Press.

Palacios, M. (2002). *Coffee in Colombia, 1850–1970: An Economic, Social and Political History*. Cambridge: Cambridge University Press.

Palacios, M. (2006). *Between Legitimacy and Violence: A History of Colombia, 1875–2002*. Durham, NC: Duke University Press.

Panhuysen, S. and Pierrot, J. (2014). *Coffee Barometer 2014*. Hivos, IUCN Netherlands, Oxfam Novib, Solidaridad and WWF.

Parra-Peña, S. R. I. (2013). *Desarrollo rural, clave para la paz en Colombia*. Decision and Policy Analysis Program – DAPA, International Center for Tropical Agriculture.

Perdomo, J. A. and Mendieta, J. C. (2007). Factores que afectan a la Eficiencia Técnica y Asignativa en el Sector Cafetero Colombiano: una aplicación con Análisis Envolvente de Datos. *Desarrollo y Sociedad*, *60*, 1–45.

Pérez, C. E. and Pérez, M. M. (2002). El sector rural en Colombia y su crisis actual. *Cuadernos de Desarrollo Rural*, *48*, 35–58.

Perez-Aleman, P. and Sandilands, M. (2008). Building value at the top and bottom of the global supply chain: MNC-NGO partnerships and sustainability. *California Management Review*, *51*(1), 24–49.

Petkova, I. (2006). Shifting regimes of governance in the coffee market: from secular crisis to a new equilibrium? *Review of International Political Economy, 13*(2), 313–339.

Pfitzer, M., Bockstette, V. and Stamp, M. (2013). Innovating for shared value. *Harvard Business Review, 91,* 100–107.

Pierrot, J., Giovannucci, D. and Kasterine, A. (2010). *Trends in the Trade of Certified Coffees.* International Trade Centre Technical Paper.

Pizano, D. (2011). *Colombian Coffee – More than 100 Years of History.* Presentation to NCA New Orleans, 17 March, Universidad de los Andes, Bogotá, Colombia.

Ponte, S. (2004). *Standards and Sustainability in the Coffee Sector. A Global Value Chain Approach.* Winnipeg: International Institute for Sustainable Development (IISD).

Portafolio (2007). *Niños, Prioridad en la Agenda Social de Nestlé.* 7 February. Retrieved from: www.portafolio.co/archivo/documento/MAM-2381375

Portafolio (2009). *Nestlé Presenta Planta de Más de US$12 Millones en Bugalagrande que Aprovecha Residuos del Café.* 4 June. Retrieved from: www.portafolio.co/archivo/documento/CMS-5357790

Portafolio (2012). *En Varias Ciudades del País el Empleo Informal Supera el 80 per cent.* 18 March. Retrieved from: www.portafolio.co/finanzas-personales/varias-ciudades-del-pais-el-empleo-informal-supera-el-80-0

Portafolio (2014). *Cómo llegar a una libra de café de US$ 134,28.* 7 September. Retrieved from: www.portafolio.co/negocios/mesa-los-santos-cafe-mas-caro-colombia

Porter, M., Hills, G., Pfitzer, M., Patscheke, S. and Hawkins, E. (n.d.). *Measuring Shared Value. How to Unlock Value by Linking Business and Social Results.* Foundation Strategy Group (FSG).

Porter, M. E. and Kramer, M. R. (2006). The link between competitive advantage and corporate social responsibility. *Harvard Business Review, 84*(12), 78–92.

Porter, M. and Kramer, M. (2011). The big idea: creating shared value. *Harvard Business Review, 89*(1), 63–77.

Posada, C. and Pontón, A. (2000). *Comercio Exterior y Actividad Económica de Colombia en el Siglo XX: Exportaciones Totales Y Tradicionales.* Bogotá: Grupo de Estudios del Crecimiento Económico Colombiano.

Potts, J., Fernandez, G. and Wunderlich, C. (2010). Prácticas comerciales para un sector cafetero sostenible: contexto, estrategias y acciones recomendadas. *Ensayos sobre Economía Cafetera,* No. 26. Federación Nacional de Cafeteros de Colombia.

PR Newswire (2007). *Sustainability Honours Mark Nespresso's AAA Sustainable Quality™ Program's Achievements and Expansion.* 18 May. Retrieved from: www.prnewswire.co.uk/news-releases/sustainability-honours-mark-nespressos-aaa-sustainable-qualitytm-programs-achievements-and-expansion-154992505.html

Presidential Council for Early Childhood – Consejería Presidencial para la Primera Infancia (2014). *La estrategia.* Retrieved from: www.deceroasiempre.gov.co

Procaña (2013). *¿Cuánto vale la tierra?* Retrieved from: www.procana.org/noticia.php?not_id=115653

Proexport Colombia (Tourism, Foreign Investment and Export Promotion) (2012). *Perfiles por Departamento Triangulo del Café.* Retrieved from: www.proexport.com.co/sites/default/files/triangulo_del_cafe.pdf.

Programa Cooperativo Regional para el Desarrollo Tecnológico y Modernización de la Caficultura/Regional Cooperative Program for the Technological Development and Modernization of Coffee Cultivation (PROMECAFÉ) (n.d.). *Acerca Promecafé.* Retrieved from: www.promecafe.org/web/

Pur Project (2014). *About Us.* Retrieved from: www.purprojet.com

Ramírez-Bacca, R. (2005). Trabajo, Familia y hacienda, Líbano-Tolima, 1923–1980. Régimen laboral-Familiar en el sistema de hacienda cafetera en Colombia. *Utopías Siglo XXI*, *3*(11): 89–98.

Ramírez-Bacca, R. (2011). Estudios e historiografía del café en Colombia, 1970–2008. Una revisión crítica. *Cuadernos de Desarrollo rural*, 7(64), 18.

Rainforest Alliance (2013). *Past Gala Honorees – Lifetime Achievement, Sustainable Standard-Setter and Green Globe Awards.* Retrieved from: www.rainforest-alliance.org/about/past-gala-honorees

Rainforest Alliance (2014) *Sustainable Agriculture.* Retrieved from: www.rainforest-alliance.org/work/agriculture/coffee

Raynolds, L. T., Murray, D. and Leigh Taylor, P. (2004). Fair Trade coffee: building producer capacity via global networks. *Journal of International Development*, *16*(8), 1109–1121.

Razeghi, A. J. (2008). *Innovating Through Recession.* Evanston, IL: Northwestern University.

Regional Economic Observatory – Observatorio Económico Regional (2013). *Bulletin.* Vol. 5, No. 13. Manizales: Universidad Autónoma de Manizales.

Report on the Outbreak of Coffee Leaf Rust in Central America and Action Plan to Combat the Pest (2013). 13 May. London: International Coffee Organisation (ICO).

Rettberg, A. (2010). Global markets, local conflict: violence in the Colombian coffee region after the breakdown of the International Coffee Agreement. *Latin American Perspectives*, *37*, 111–132.

Rice, P. and McLean, J. (1999). *Sustainable Coffee at the Crossroads.* A White Paper prepared for the Consumers Choice Council.

Riveros Saveedra, A. (2013). *Crecimiento Económico y Conflicto Interno en Colombia.* Grupo Bancolombia.

Rivillas Osorio, C., Serna, C., Cristancho, M. and Gaytán, A. (2011). *La Roya del Cafeto en Colombia. Impacto, Manejo y Costos del Control.* Caldas: Centro Nacional de Investigaciones de Café (Cenicafé).

RobecoSAM AG (2013). *Industry Group Leader Report / 2013 Nestle SA.* RobecoSAM AG. Sustainability Indices.

Rodríguez, D. and Duque, A. (2009). El paisaje cultural cafetero: reflexiones desde la diversidad agrícola y las percepciones históricas de la naturaleza y la cultura. *Diálogo entre saberes: ciencias e ideologías en torno a lo ambiental.* Pereira, Universidad Tecnológica de Pereira, 121–128.

Rodrik, D. (2008). *Coffee Prices, Oil Prices, and Violence in Colombia.* Retrieved from: http://berkeleyinternationalpolicy.wordpress.com/2008/10/23/coffee-prices-oil-prices-and-violence-in-colombia/

Roldán-Pérez, A., Gonzalez-Perez, M. A., Huong, T., Tien, D. N., Riegler, F. X., Riegler, S., ... and City, H. C. M. (2009). Coffee, cooperation and competition: a comparative study of Colombia and Vietnam. *UNCTAD. UNCTAD Virtual Institute.*

Roseberry, W., Gudmundson, L. and Kutschbach, M. (1995). *Coffee, Society, and Power in Latin America.* Baltimore: Johns Hopkins University Press.

Ruben, R. and Zuniga, G. (2011). How standards compete: comparative impact of coffee certification schemes in northern Nicaragua. *Supply Chain Management: An International Journal*, *16*(2), 98–109.

Rueda, X. and Lambin, E. F. (2013). Responding to globalization: impacts of certification on Colombian small-scale coffee growers. *Ecology and Society*, *18*(3), 21.

Sanz, C. G. C., Mejía, C. V., García, E. C., Torres, J. S. A. and Calderón, E. Y. T. (2012). *El Mercado Mundial del Café y su Impacto en Colombia* (No. 009612). Banco de la República.

Sector de la Minería a Gran Escala (SMGE) (2013). *El sector minero en Colombia: Impactos macroeconómicos y encadenamientos sectoriales.* Fundación para la Educación Superior y el Desarrollo (Fedesarrollo).

Semana, Rumbo Pacífico (2013). *Publicaciones Semana, Bogotá.* Retrieved from: www.semana.com/especiales/Semana__Rumbo_Pacfico/index.html#/2/

Sieber, N. (2008). *The World Development Report 2008: Freight Transport for Development: A Policy Toolkit.* Washington, DC: The World Bank.

Silva, G. (2004). *Organizaciones Privadas, Dividendos Públicos.* Federación Nacional de Cafeteros, Ensayos No. 22.

Silverman, J. (2007). *Proyecto Multinacionales y Medioambiente Informe: Colombia.* Medellín: Escuela Nacional Sindical.

Sindicato Nacional de Trabajadores de la Industria de Alimentos (SINALTRAINAL) (n.d.). *Caso 4 Contaminación Hídrica Generada por Nestlé en Colombia.* NIT. 860.517.322-7. Diario Oficial No. 36207.

Sintercafé (2014). "About Us." Retrieved from: www.sintercafe.com/en/pages/about_us

Situación de Buenos Precios del Café Podría Ser para Largo (2014). Retrieved from: www.elnuevodia.com.co/nuevodia/actualidad/economica/211081-situacion-de-buenos-precios-del-cafe-podria-ser-para-largo#sthash.wdiPClK3.dpuf

Sullivan, R. (2003). Human rights and companies in conflict zones. *Conflict, Security & Development, 3*(2), 287–299.

SustainAbility, United Nations Environment Programme and United Nations Global Compact (2008). *Unchaining Value. Innovative Approaches to Sustainable Supply.* Retrieved from: www.unglobalcompact.org/docs/news_events/8.1/unchaining_value.pdf

Sustainability Education Network (2010) *Welcome.* Retrieved from: http://sen4earth.org

Sustainability MBA Challenge (2014). *MBA Challenge 2014.* Retrieved from: http://sustainabilitymbachallenge.com/the-challenge/

Sustainable Agriculture Initiative (SAI). (2013). *Sustainable Agriculture Initiative Annual Report 2013.* Sustainable Agriculture Initiative Platform.

Sustainable Agriculture Network (2010). *SAN Principles.* Retrieved from: http://sanstandards.org/sitio/subsections/display/7

Sustainable Coffee Project (2014). "About". Retrieved from: www.sustainablecoffeeprogram.com/en/about-coffee

Sustainable Markets Intelligence Centre (CIMS) (2013). *MBA Students Rise to the Challenge.* 10 June. Retrieved from: /www.cims-la.com/blog/mba-students-rise-challenge

TechnoServe (2009). *Creating Shared Value Through Private Public Partnerships. A case study look at the TechnoServe-Nespresso effort in Caldas, Colombia 2006–2008.* Retrieved from: www.syngentafoundation.com/content/api/org_files/nespresso_technoserve_ppp_final.pdf

The Blade (1980). *Swiss Chemist Known as 'Father of Nescafé Instant Coffee'.* 11 September. The Blade Home Edition, p. 15. Toledo, Ohio.

The Green Organisation (2012). *Nespresso Meets Recycling Commitment One Year Ahead.* 3 October. Retrieved from: www.thegreenorganisation.info/nespresso-meets-recycling-commitment-one-year-ahead/

Thorp, R. (2000). *Has the Coffee Federation become Redundant? Collective Action and the Market in Colombian Development.* Helsinki: International Coffee Organisation (ICO).

Toro, G. (2005). Eje Cafetero colombiano: compleja historia de caficultura, violencia y desplazamiento. *Revista de Ciencias Humanas, 35,* 127–149. Manizales: Universidad de Caldas.

Township of Bugalagrande-Valle (2010). *Nuestro Municipio.* Retrieved from: http://bugalagrande-valle.gov.co

United Nations (UN) (2009). The coffee trade and its MDG ramifications. *UN Chronicle,* *44*(4).

United Nations Development Programme (UNDP) (2011). *Desarrollo rural con enfoque territorial: desafío para la política pública.* Retrieved from: www.pnud.org.co/hechosdepaz/64/desarrollo_territorial_con_enfoque_territorial.pdf

United Nations Development Programme (UNDP) (2014). *Green Commodities Programme.* Retrieved from: www.undp.org/content/undp/en/home/ourwork/environmentand energy/projects_and_initiatives/green-commodities-programme/

United Nations Environment Programme (UNEP) (2012). *Sustainable Supply Chains Allow SMEs to Reap the Benefits of the Green Economy Transition.* Colombia: Federación Nacional de Cafeteros.

United Nations Global Compact (2010). *Supply Chain Sustainability. A Practical Guide for Continuous Improvement.* New York: UN.

United Nations Global Compact (2013). Global Corporate Sustainability Report 2013. *UN Global Compact Reports, 5*(1), 1–28.

United Nations Global Compact (2014). *Corporate Sustainability in the World Economy.* New York: UN.

United Nations Office for the Coordination of Humanitarian Affairs (OCHA) (2014). *Colombia: Impacto humanitario de Paro Agrario. Flash Update No. 1.* 7 May. Retrieved from: http://reliefweb.int/sites/reliefweb.int/files/resources/140507 per cent20Flash per cent20Update per cent20impacto per cent20paro per cent20agrario.pdf

United States Agency for International Development (USAID) (1982). *Rural Roads Evaluation Summary Report.* A.I.D. Program Evaluation Report No 5. USAID.

United States Agency for International Development (USAID) (2010). *Audit of USAID/ Colombia's Alternative Development Program* (Audit Report No. 1-514-10-004-P.), San Salvador, El Salvador.

Valkila, J. (2009). Fair Trade organic coffee production in Nicaragua – sustainable development or a poverty trap? *Ecological Economics, 68*(12), 3018–3025.

Vallecilla, G. (2005). Cien Años del Café en Caldas. *Estudios Regionales, CRECE, Manizales,* 16.

Vargas, J. F. (2003). *Conflicto Interno y Crecimiento Económico en Colombia.* (Tesis Maestría en Economía). Colombia: Universidad de los Andes.

Wile, R. (2014). Why coffee prices are exploding. *Business Insider Singapore.* Retrieved from: www.businessinsider.sg/why-coffee-prices-are-surging-2014-2/#.UxHlxlz8pzQ

World Coffee Research (2013). *First International Coffee Rust Summit: Final Report, Guatemala, April 18–20.* World Coffee Research Annual Report 2012-2013.

World Coffee Research and PROMECAFE (2013). *The Coffee Crisis in Mesoamerica. Causes and Appropriate Responses.* Inter-American Institute for Cooperation on Agriculture (IICA).

World Economic Forum (WEF) (2009). *The Business Case for Sustainability.* Cologne: WEF.

World Economic Forum (WEF) (2014). *Sustainable Consumption.* Retrieved from: www.weforum.org/issues/sustainable-consumption

World Heritage Centre (2008). *Operational Guidelines for the Implementation of the World Heritage Convention.* UNESCO. Paris. Retrieved from: http:/whc.unesco. org/en/guidelines.

Yang, J. L. (2014). Don't panic. But there's a global coffee shortage. *Washington Post,* 21 February. Retrieved from: www.washingtonpost.com/blogs/wonkblog/wp/2014/02/21/dont-panic-but-theres-a-global-coffee-shortage/

Zecuppa Coffee (n.d.). *Glossary of Coffee Terminology.* Retrieved from: www.zecuppa.com/coffeeterms-bean-grading.htm

Index

For Product Safety Concerns and Information please contact our
EU representative GPSR@taylorandfrancis.com Taylor & Francis
Verlag GmbH, Kaufingerstraße 24, 80331 München, Germany